JUXTAPOSITION

Irina Petrova Adamatzky

JUXTAPOSITION

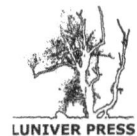

LUNIVER PRESS

Published by Luniver Press Bristol BS39 5RX United Kingdom

British Library Cataloguing-in-Publication Data
A catalog record for this book is available from the British Library

Juxtaposition. 2022

Copyright © Luniver Press 2022

All rights reserved. This book, or parts thereof, may not be reproduced in any form or by any means, electronic or mechanical, including photocopying, recording or by any information storage and retrieval system, without permission in writing from the copyright holder.

ISBN-10: 1-905986-13-0
ISBN-13: 978-1-905986-13-2

While every attempt is made to ensure that the information in this publication is correct, no liability can be accepted by the authors or the publishers for loss, damage or injury caused by any errors in, or omission from, the information given.

Dedicated to my husband Andrew Adamatzky,
who inspired me to create this book.

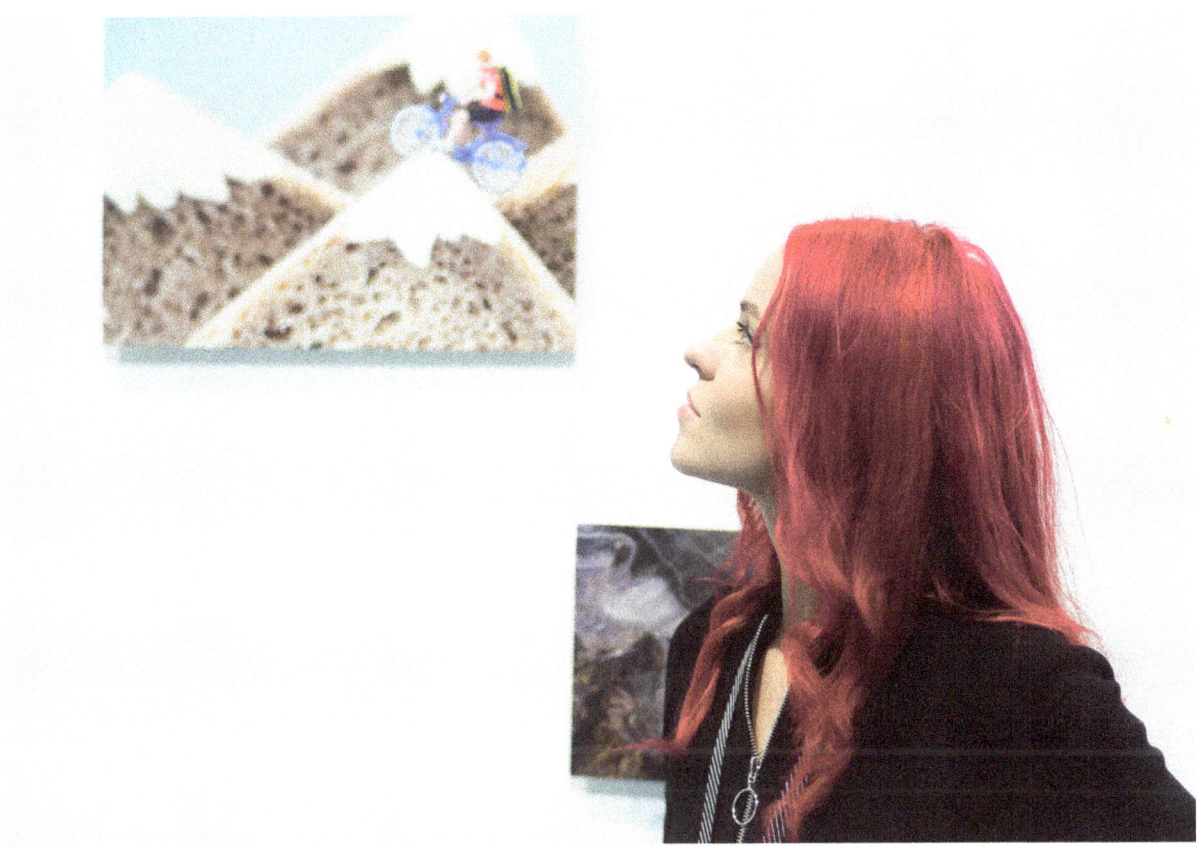

Irina Petrova Adamatzky is an award winning and published artist and photographer. She specializes in wild life micro-photography and science fiction inspired installations organically integrating living and artificial entities. A unique feature of her work is she uses retro manual focus lenses to share the wonders of the world.

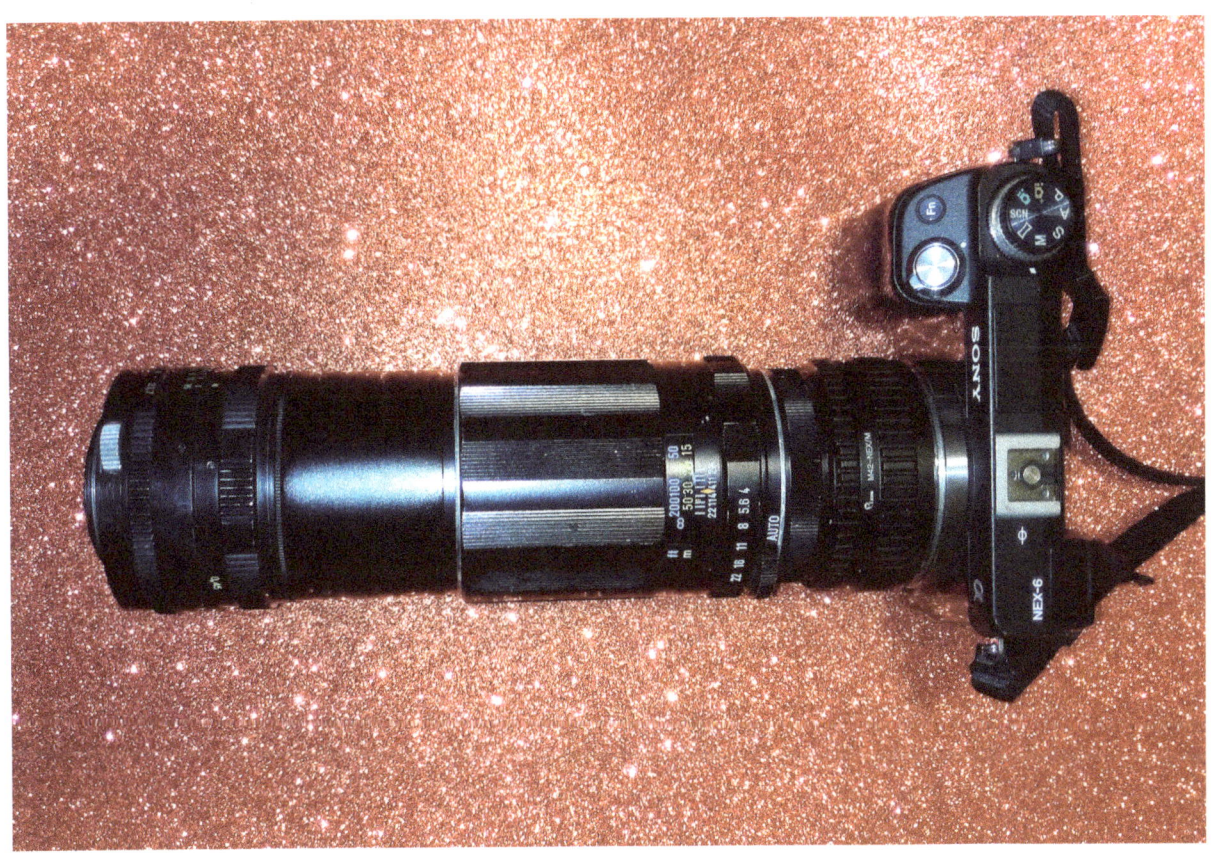

I started this project in 2018 to show that the unusual can be found anywhere, even in such a cold place as the Southern Ural, where I lived at that time. All I had at my disposal was an old camera SONY Alpha NEX-6 and a few low-cost retro lenses, which I connected with an adapter to achieve high magnification.

Irina Petrova Adamatzky

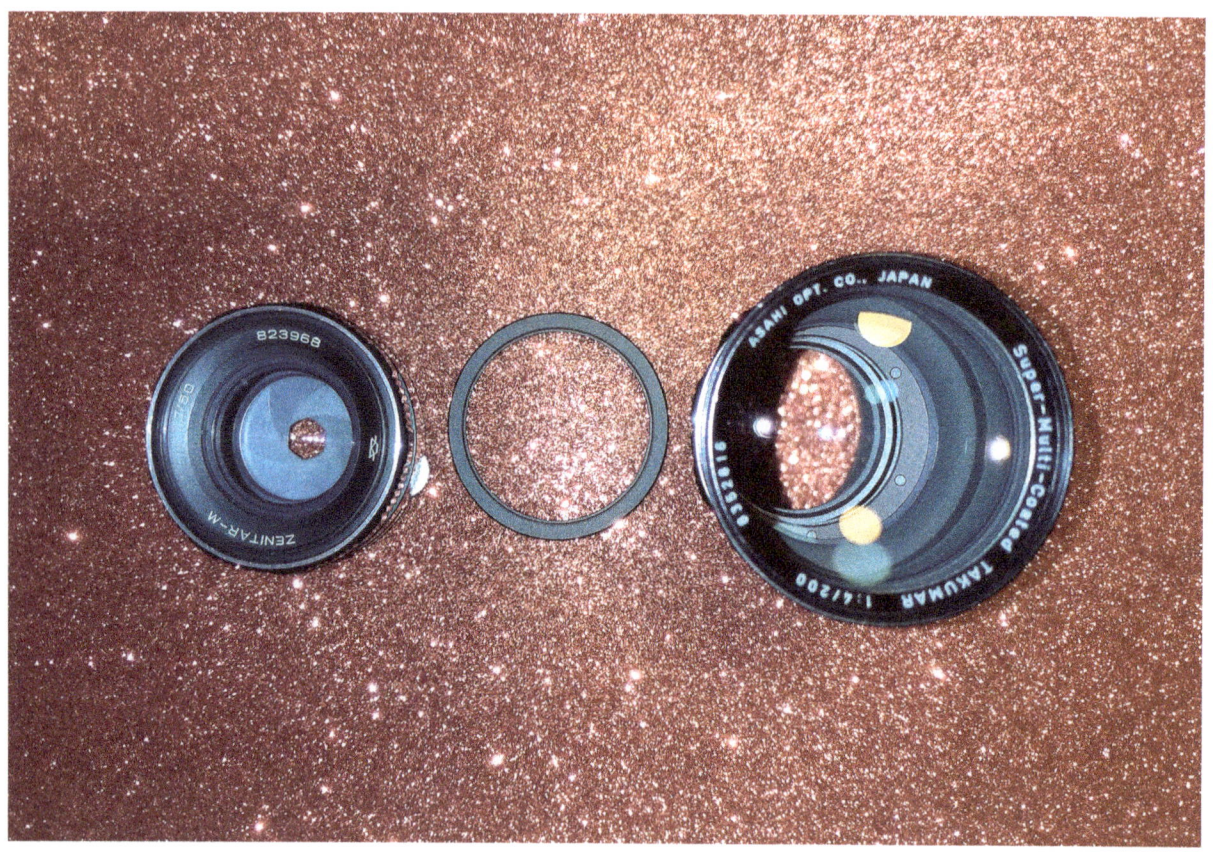

I had no idea what a huge and wonderful world you could discover with this kind of equipment. This project completely changed me and my attitude to the microcosm under my feet. Hope it helps people see beauty in ordinary little things.

Irina Petrova Adamatzky, Bristol, January 2022

Juxtaposition

Irina Petrova Adamatzky

Juxtaposition

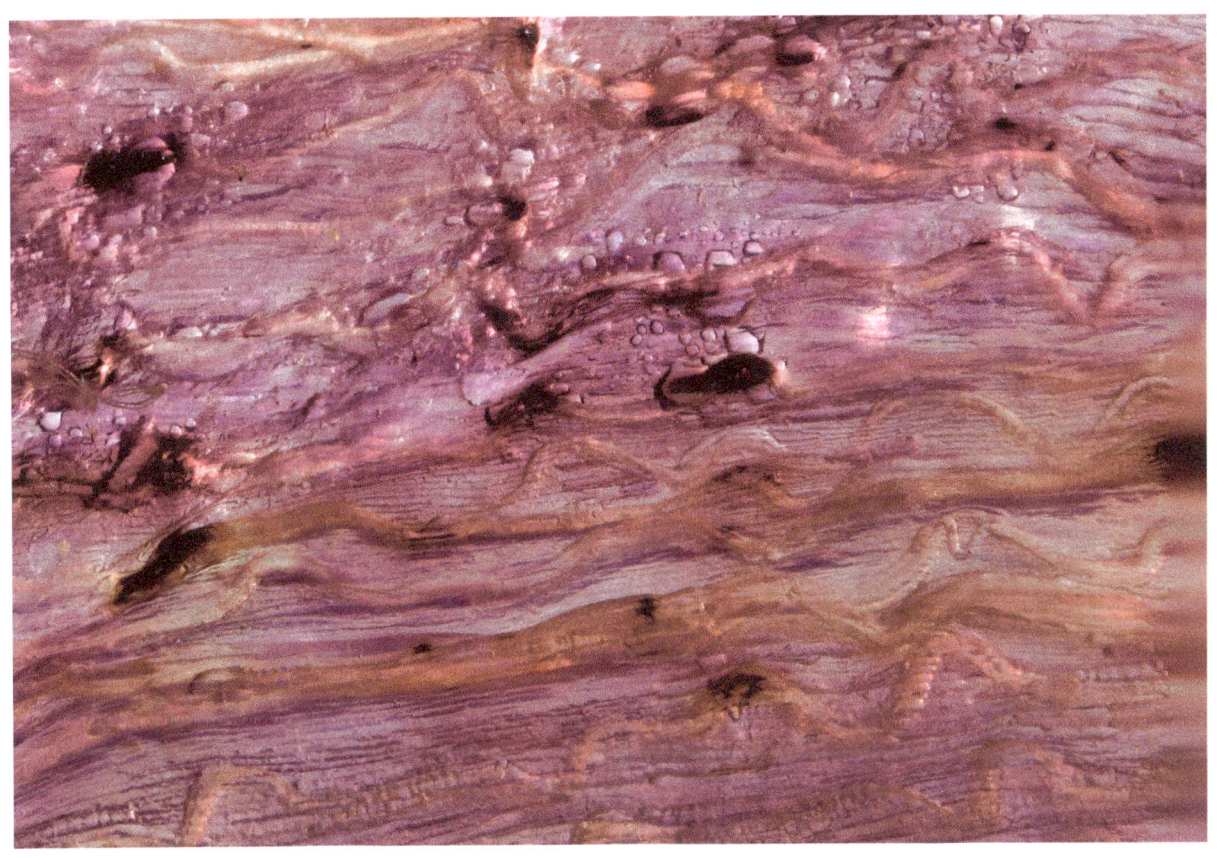

A photo of a tulip (*Tulipa*) petal taken with two lenses as one lens: MC Jupiter-37A 135mm f/3,5, which was made in 1985, and Zenitar-M 1,7/50, made in 1982. Magnification is 2.5x.

Irina Petrova Adamatzky

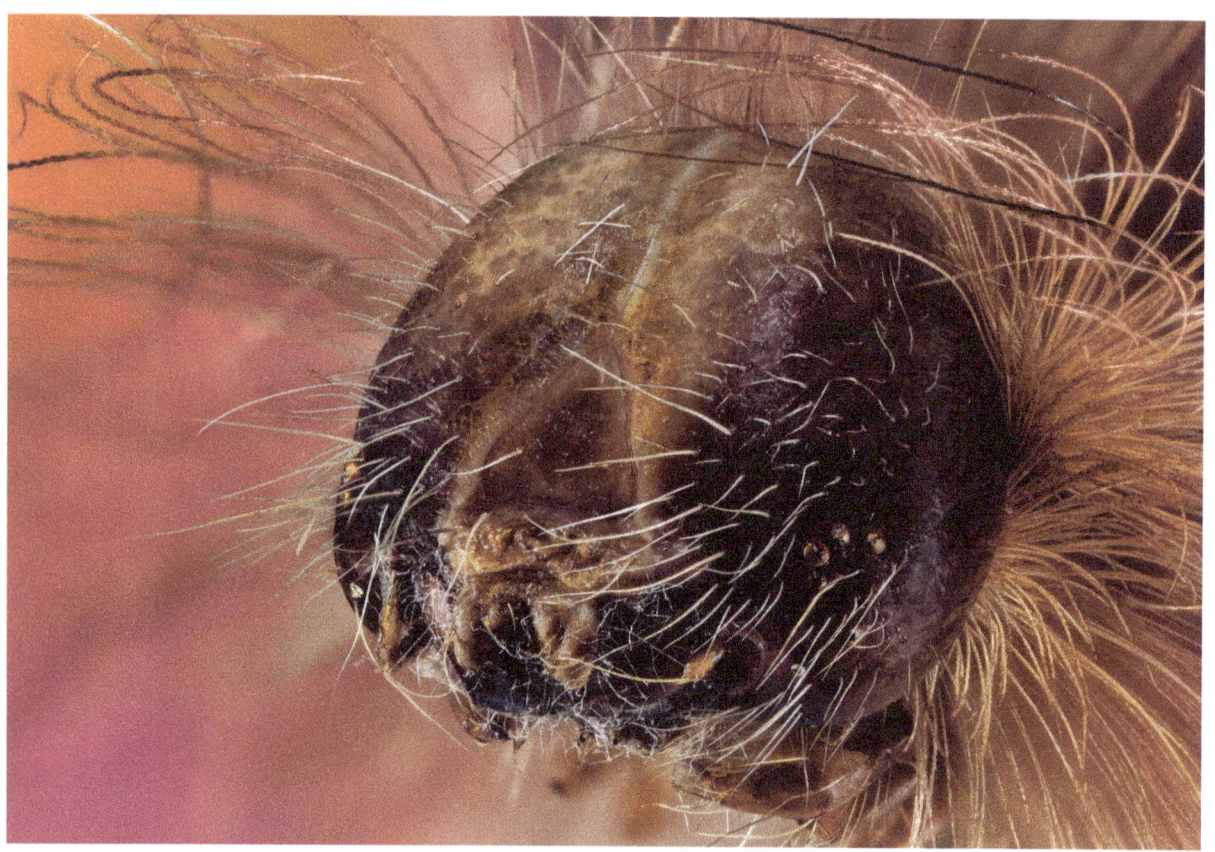

A microcosmic portrait of a pale tussock (*Calliteara pudibunda*) caterpillar taken with two lenses as one lens: MC Jupiter-37A, which was made in 1985, and Zenitar-M 1,7/50, made in 1982. This image was stacked from 225 shots into one very sharp photo. Magnification is 2,5x.

Juxtaposition

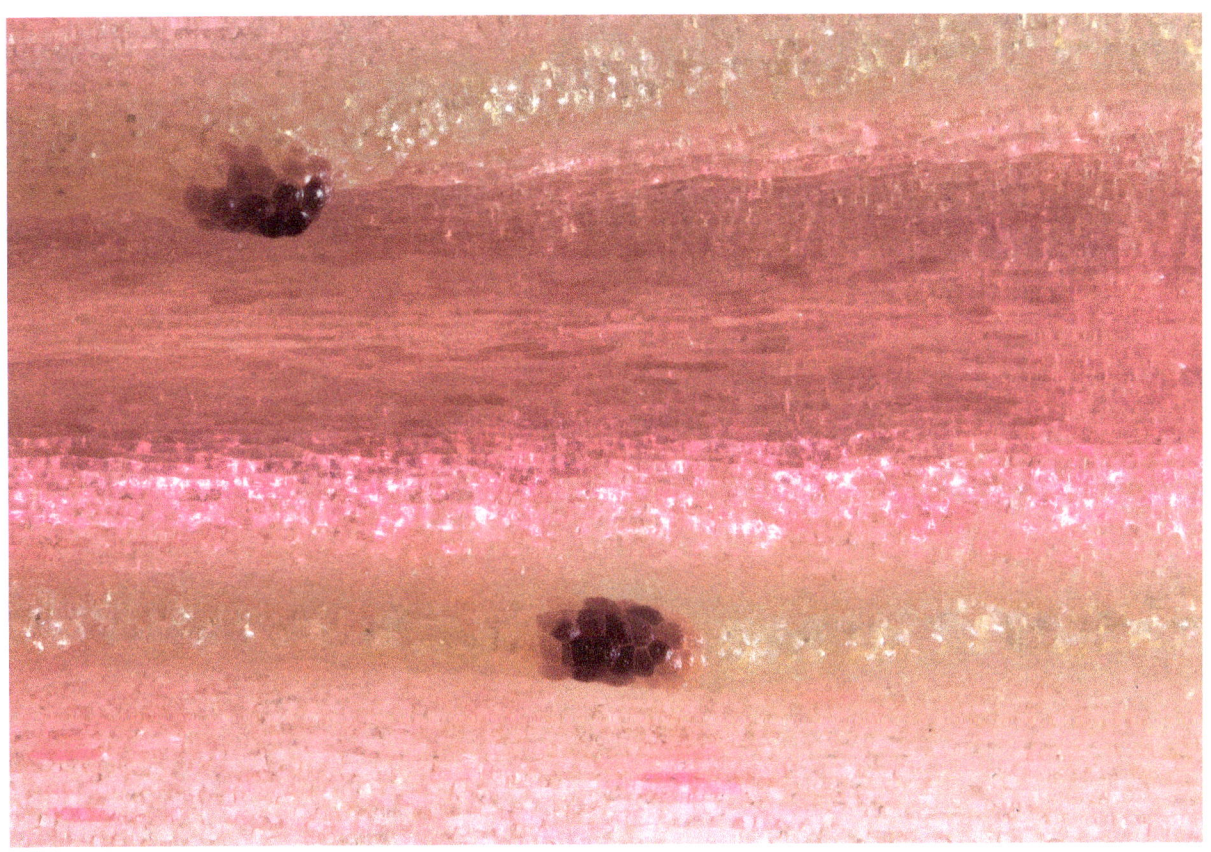

A micro photo of a lily (*Lilium*) petal taken with two lenses as one lens: SMC Takumar 200mm f/4, which was made in 1982, and Zenitar-M 1,7/50, made in 1982. This image was stacked from many shots into one very sharp photo. Magnification is 4x.

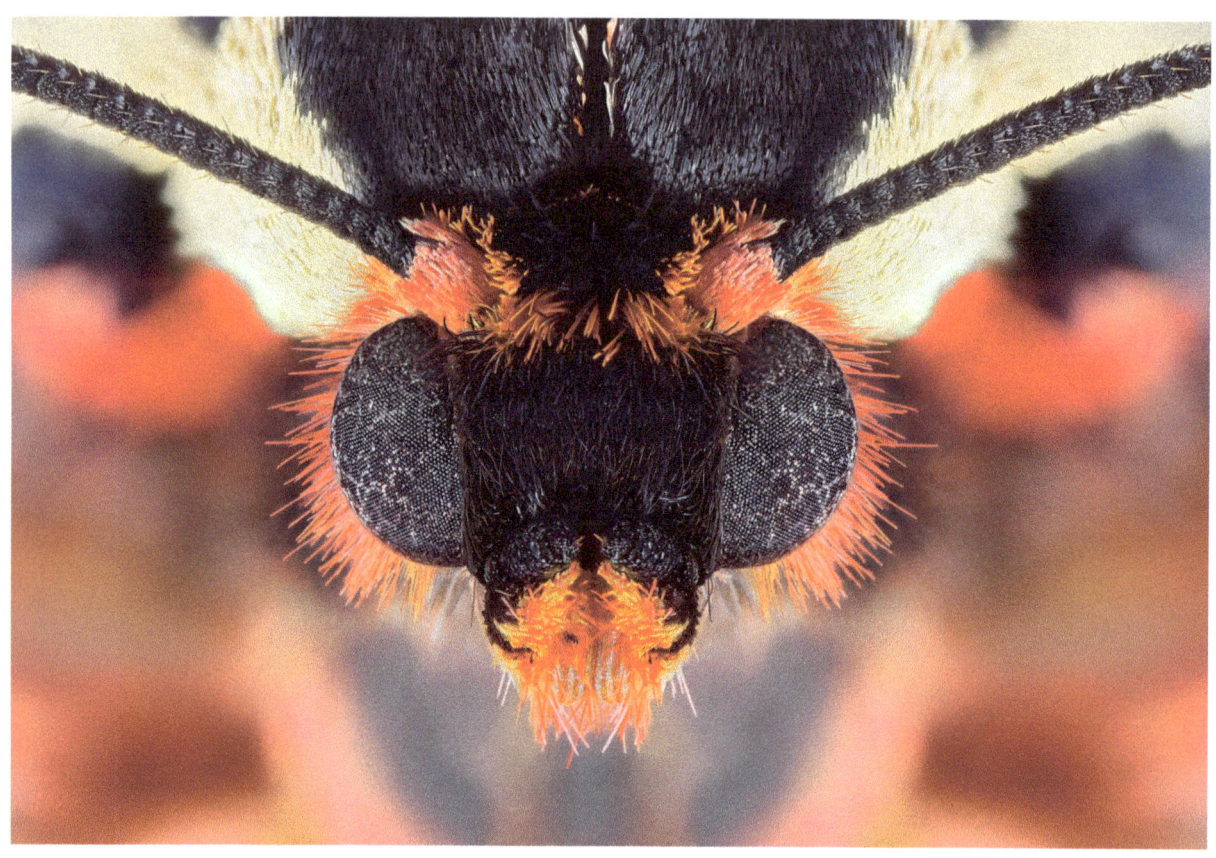

A microcosmic portrait of a garden tiger moth (*Arctia caja*) taken with two lenses as one lens: MC Jupiter-37A, which was made in 1985, and Zenitar-M 1,7/50, made in 1982. This image was stacked from 46 shots into one very sharp photo. Magnification is 2,5x.

Juxtaposition

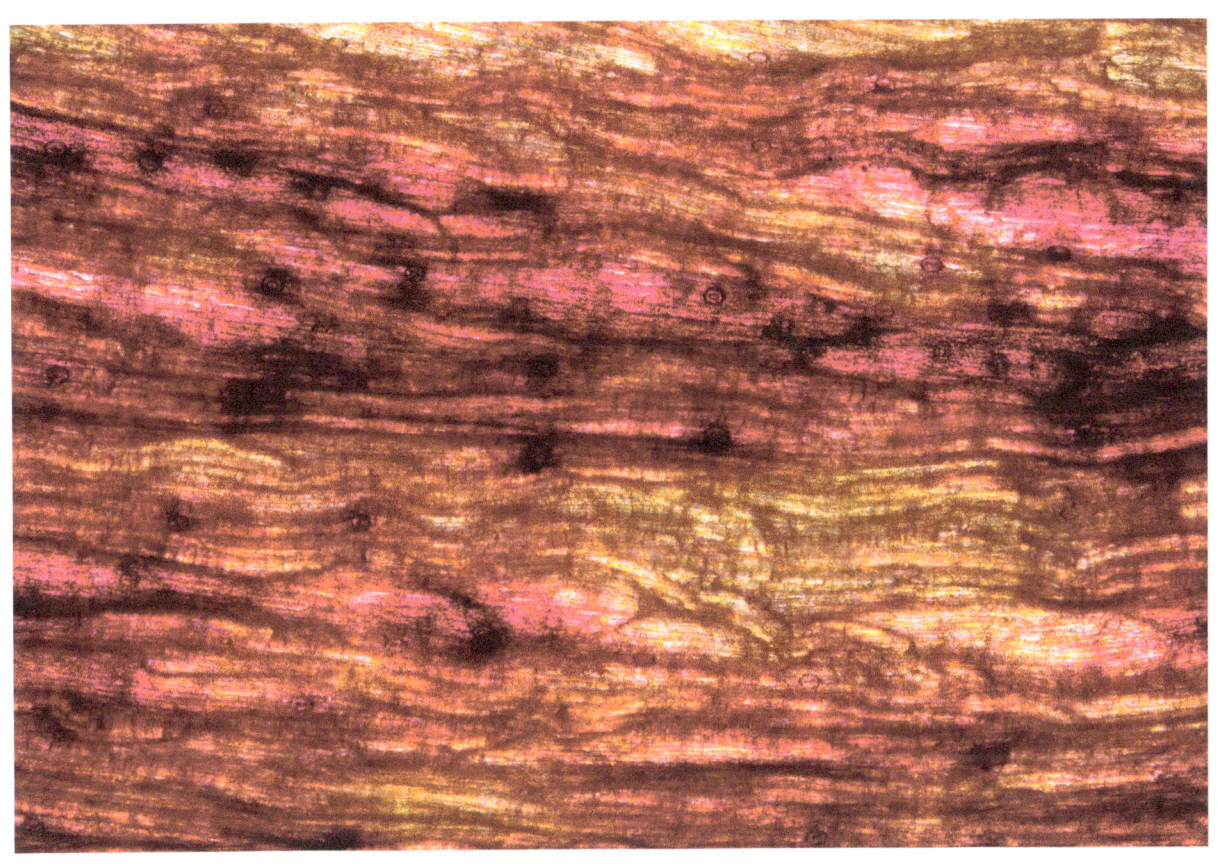

A photo of a tulip (*Tulipa*) petal taken with two lenses as one lens: MC Jupiter-37A 135mm f/3,5, which was made in 1985, and Zenitar-M 1,7/50, made in 1982. Magnification is 2.5x.

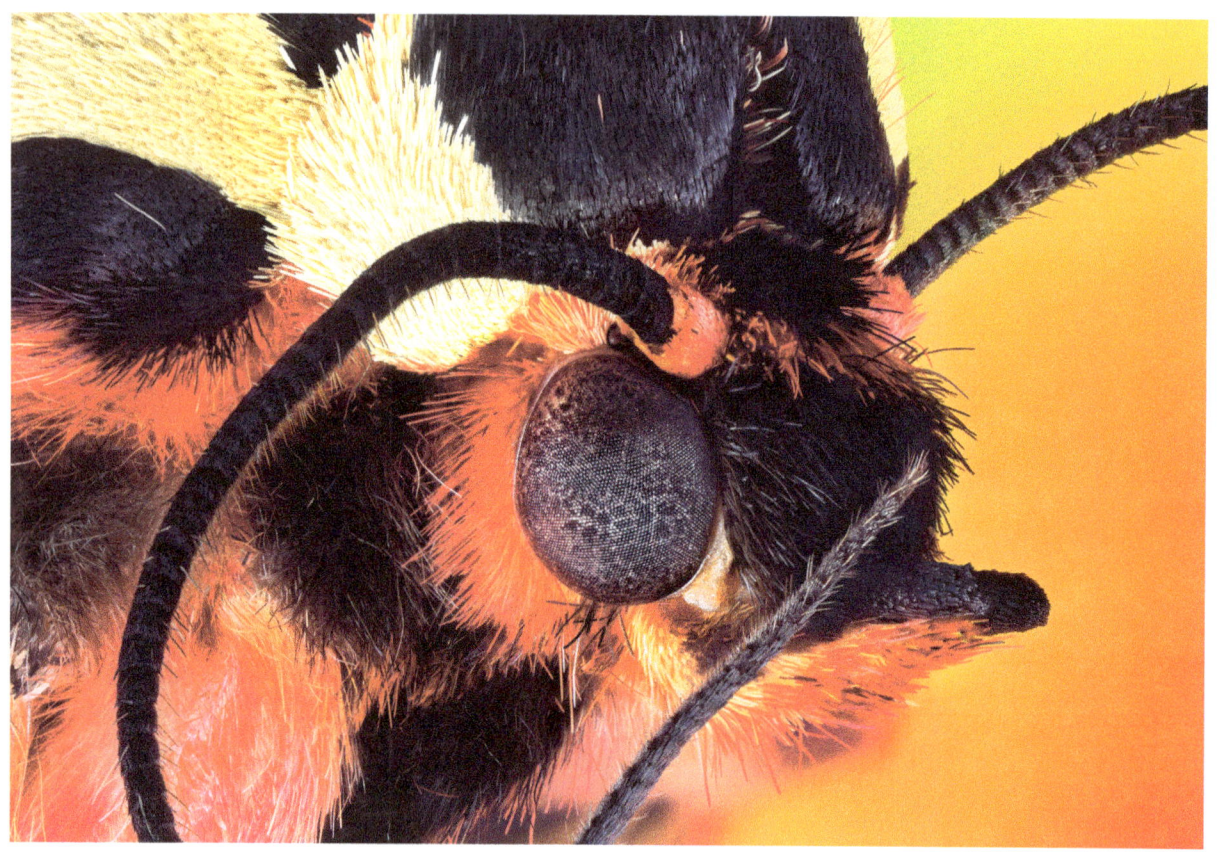

A microcosmic portrait of a garden tiger moth (*Arctia caja*) taken with two lenses as one lens: MC Jupiter-37A, which was made in 1985, and Zenitar-M 1,7/50, made in 1982. This image was stacked from 46 shots into one very sharp photo. Magnification is 2,5x.

Juxtaposition

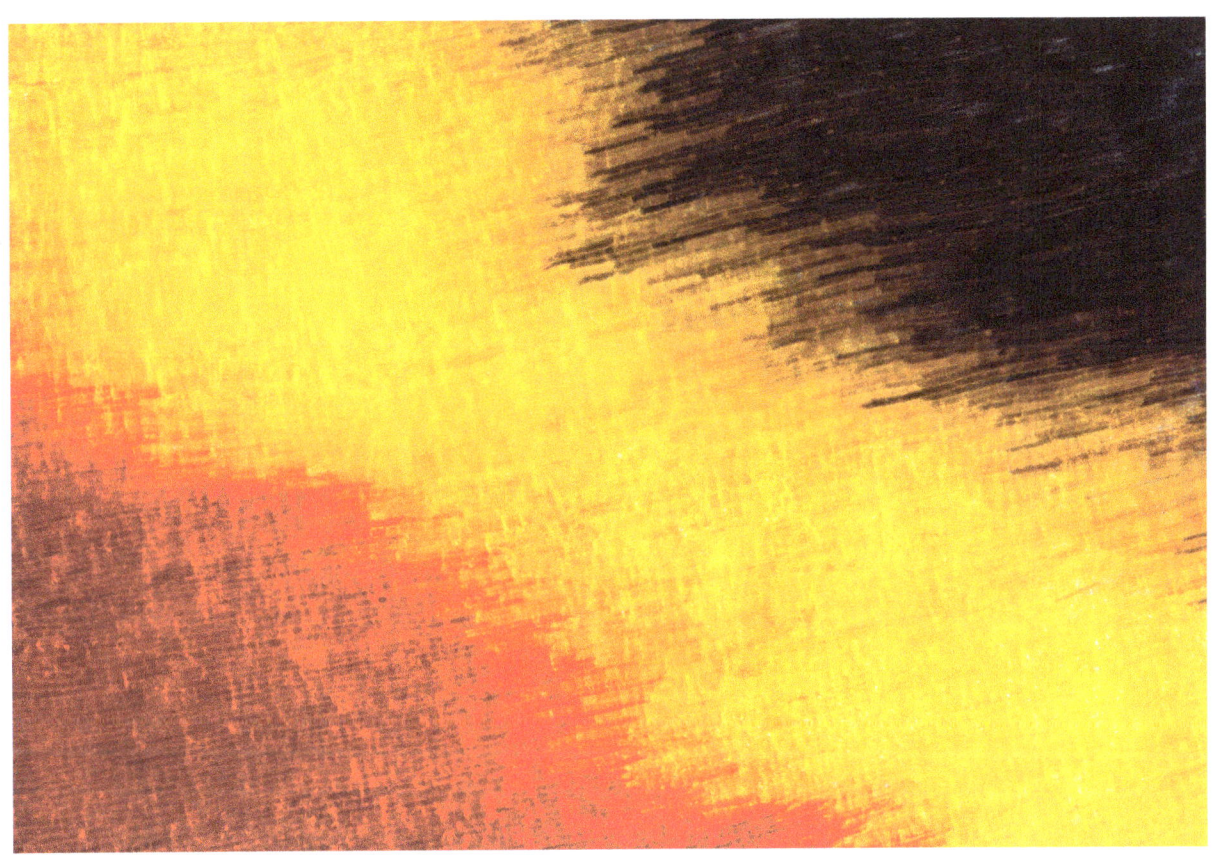

A photo of a tulip (*Tulipa*) petal taken with two lenses as one lens: MC Jupiter-37A 135mm f/3,5, which was made in 1985, and Zenitar-M 1,7/50, made in 1982. Magnification is 2.5x.

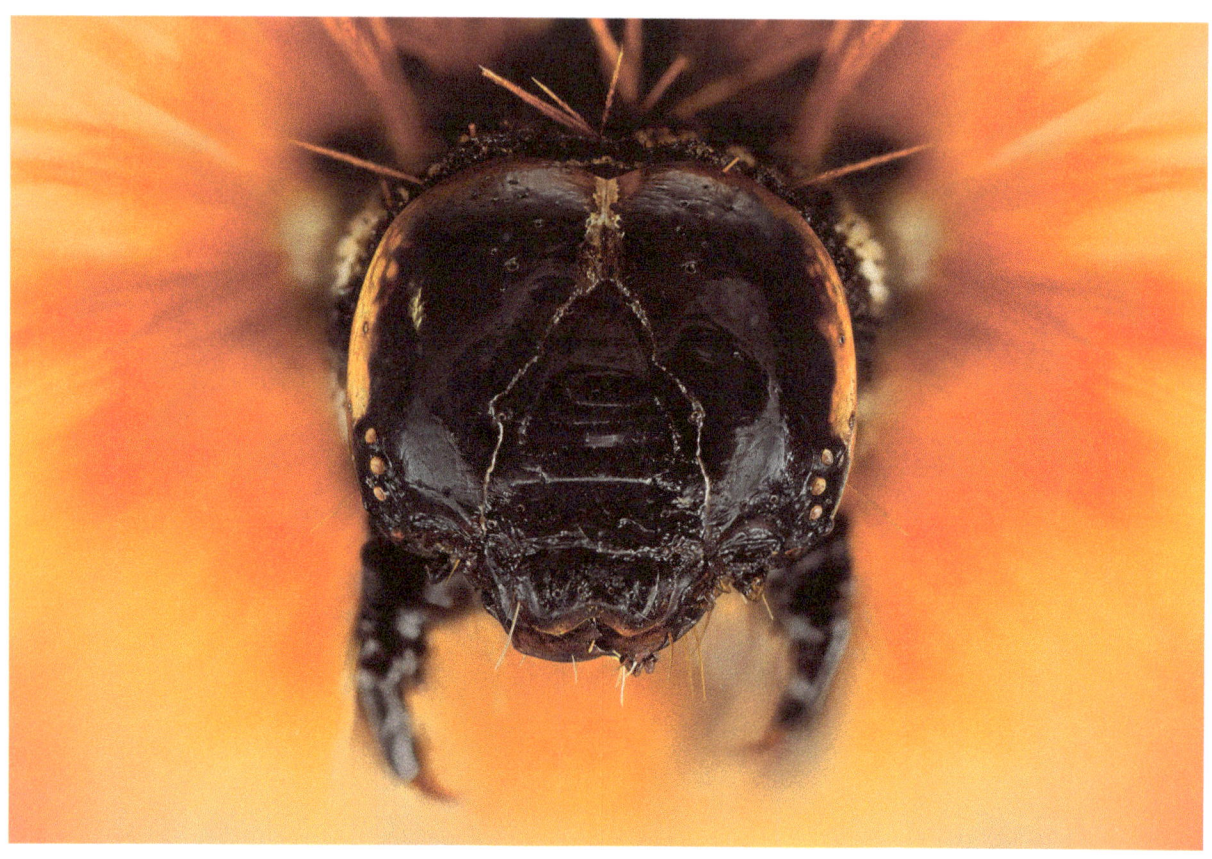

A microcosmic portrait of a caterpillar of a garden tiger moth (*Arctia caja*) taken with two lenses as one lens: MC Jupiter-37A, which was made in 1985, and Zenitar-M 1,7/50, made in 1982. This image was stacked from 203 shots into one very sharp photo. Magnification is 2,5x.

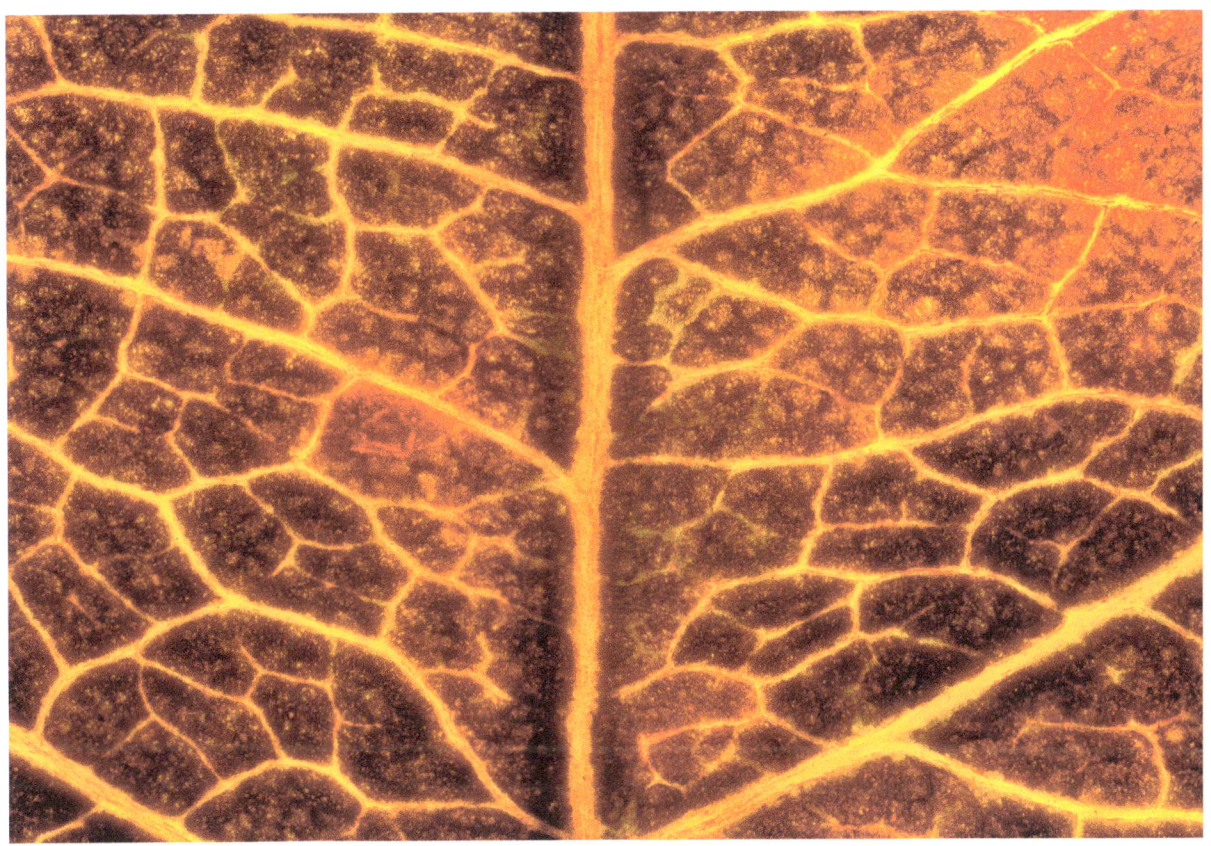

A photo of a birch (*Betula*) leaf taken with two lenses as one lens: MC Jupiter-37A 135mm f/3,5, which was made in 1985, and Zenitar-M 1,7/50, made in 1982. This image was stacked from many shots into one very sharp photo.
Magnification is 2,5x.

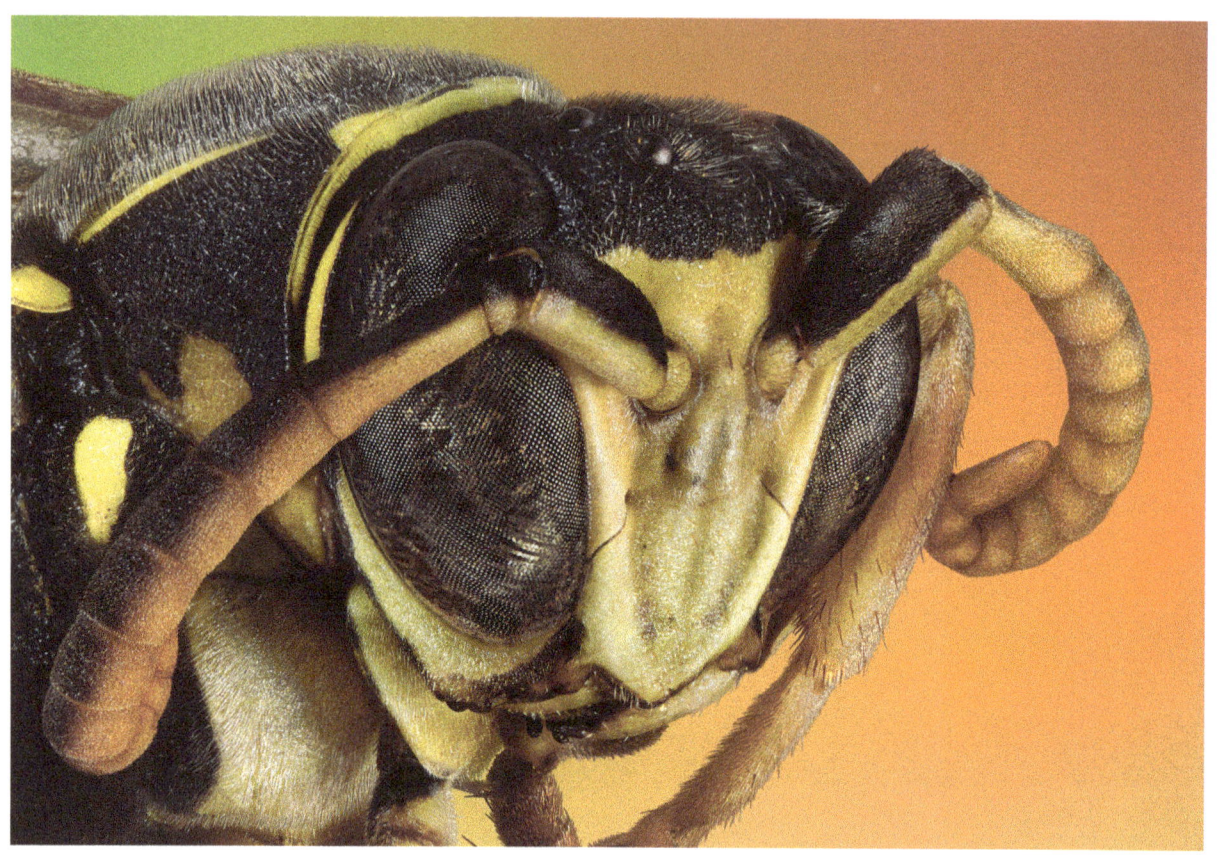

A microcosmic portrait of a paper wasp (*Polistinae*) taken with two lenses as one lens: MC Jupiter-37A, which was made in 1985, and Zenitar-M 1,7/50, made in 1982. This image was stacked from 240 shots into one very sharp photo. Magnification is 2,5x.

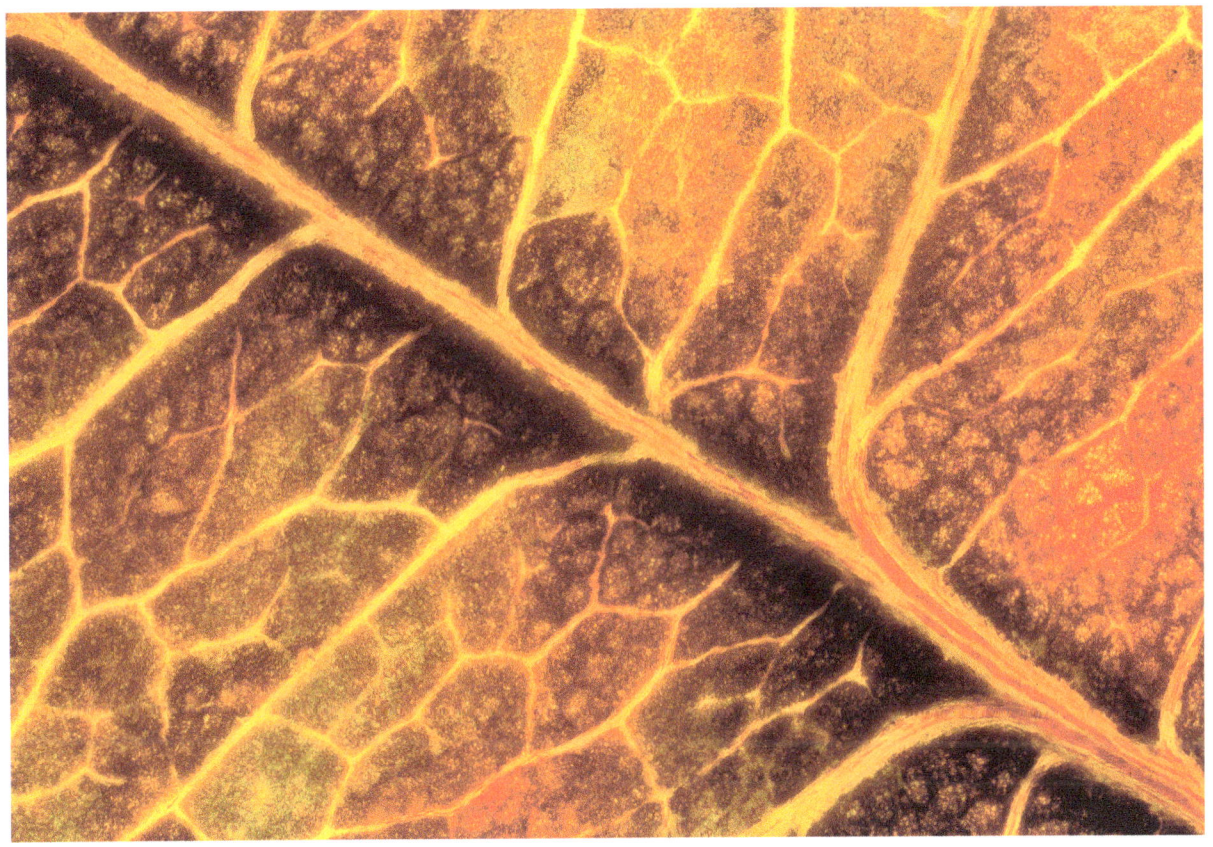

A photo of a birch (*Betula*) leaf taken with two lenses as one lens: MC Jupiter-37A 135mm f/3,5, which was made in 1985, and Zenitar-M 1,7/50, made in 1982. This image was stacked from many shots into one very sharp photo.
Magnification is 2,5x.

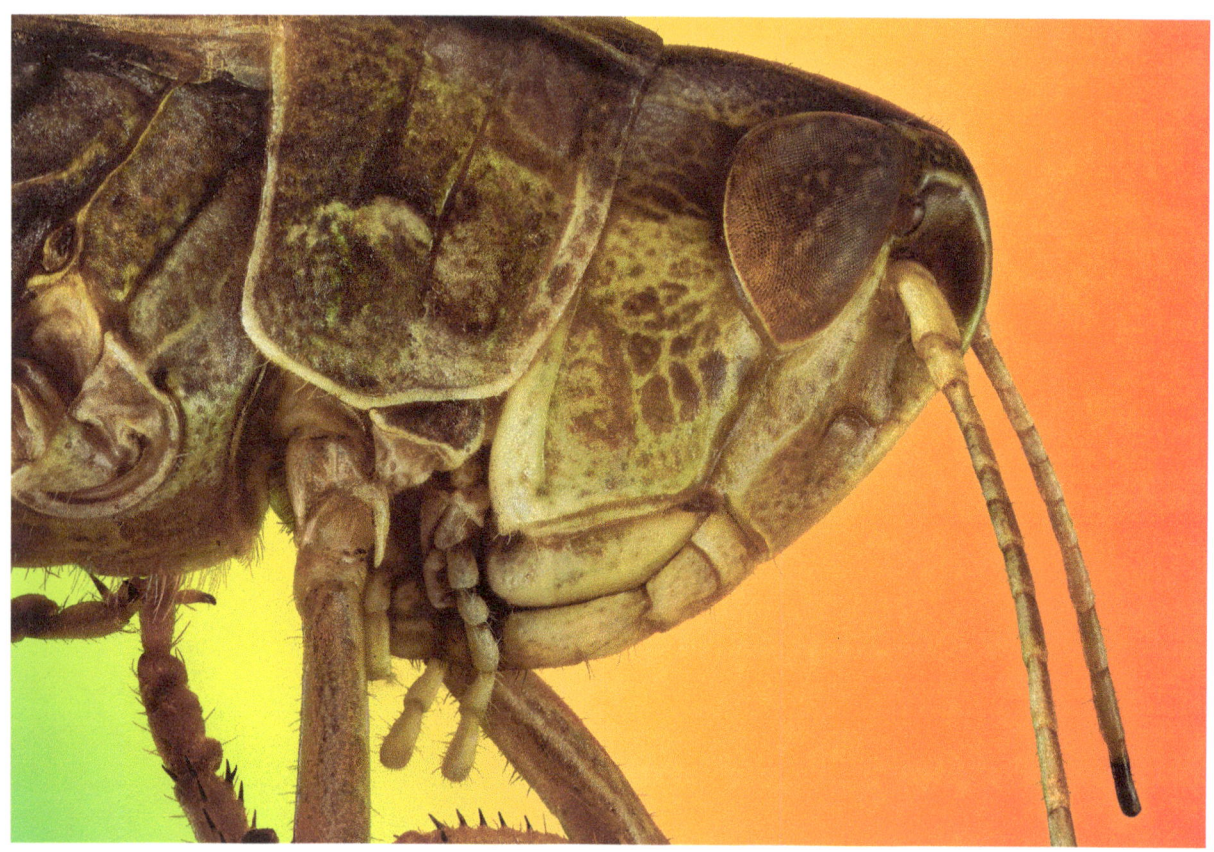

A microcosmic portrait of a locust (*Gomphocerus sibiricus*) taken with two lenses as one lens: MC Jupiter-37A, which was made in 1985, and Zenitar-M 1,7/50, made in 1982. This image was stacked from 150 shots into one very sharp photo. Magnification is 2,5x.

Juxtaposition

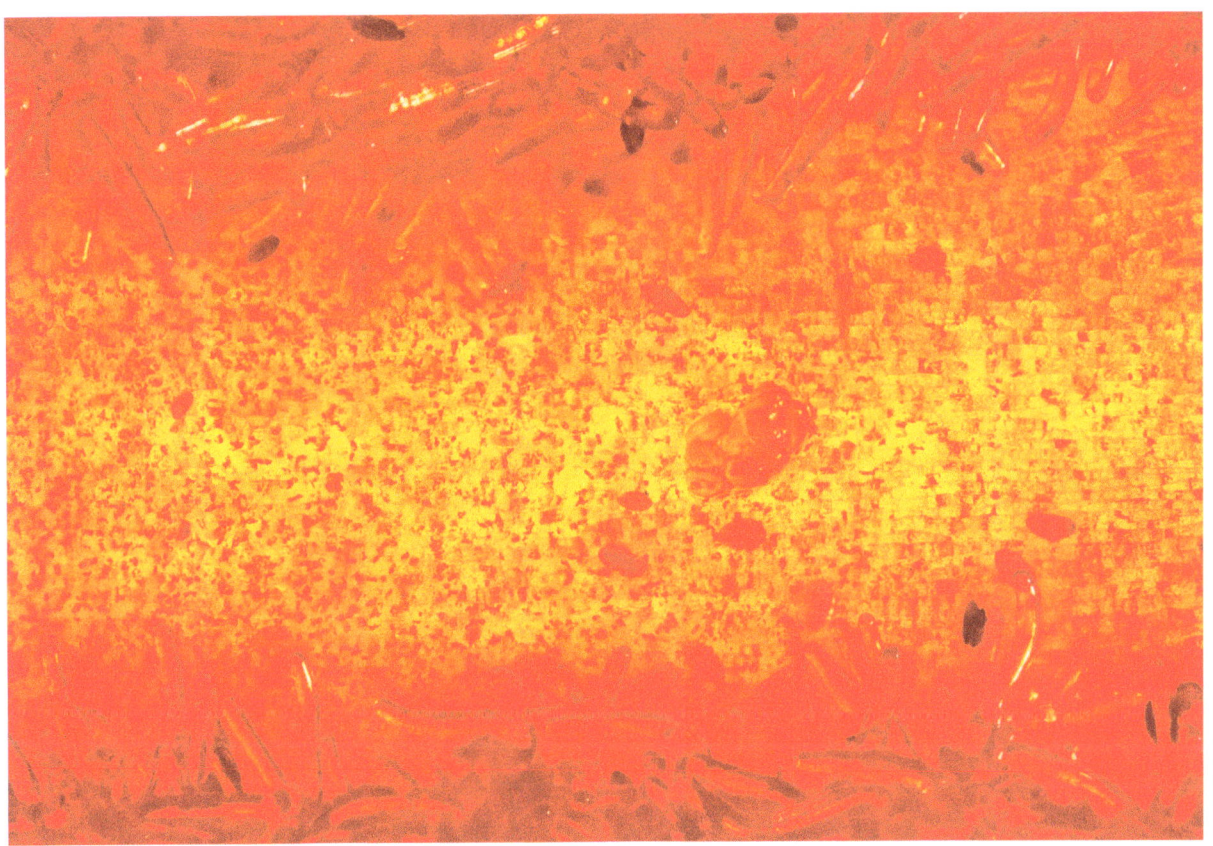

A micro photo of a lily (*Lilium*) petal taken with two lenses as one lens: SMC Takumar 200mm f/4, which was made in 1982, and Zenitar-M 1,7/50, made in 1982. This image was stacked from many shots into one very sharp photo. Magnification is 4x.

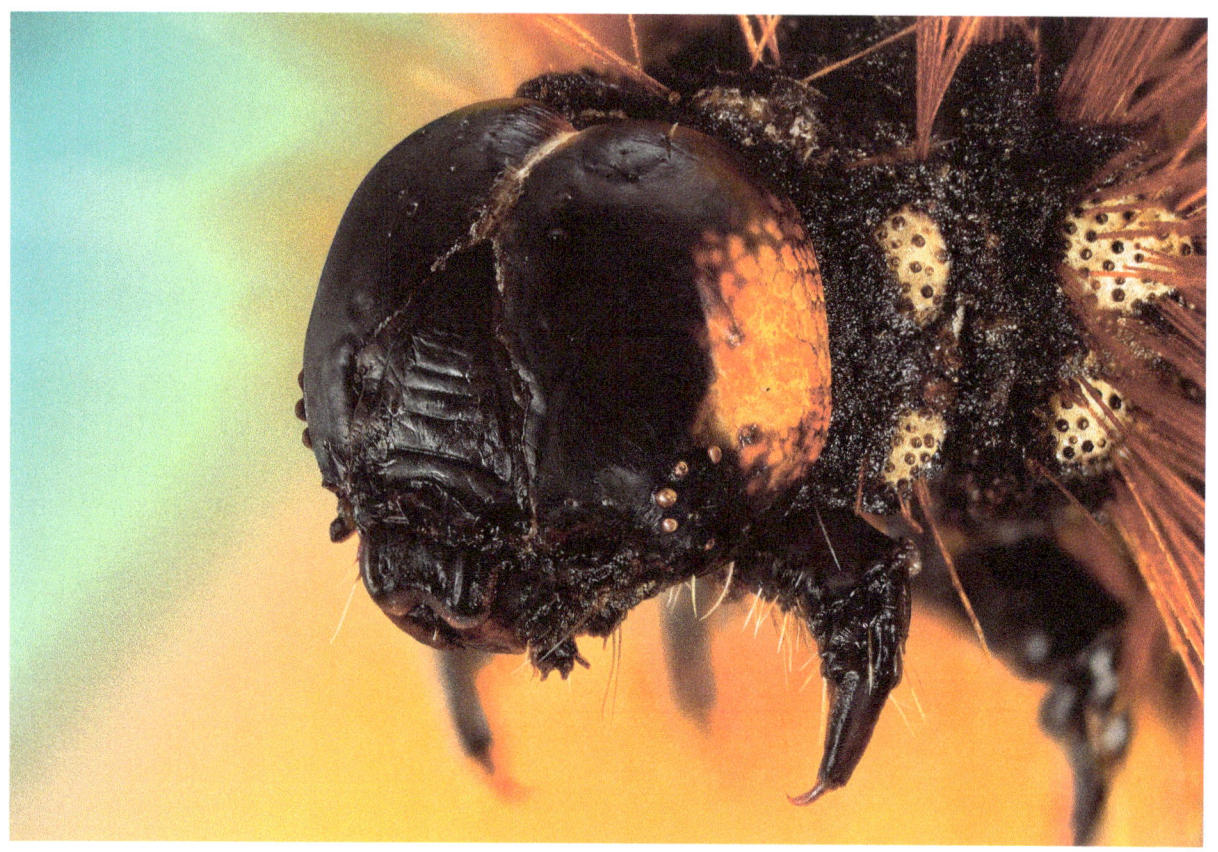

A microcosmic portrait of a garden tiger moth (*Arctia caja*) caterpillar taken with two lenses as one lens: MC Jupiter-37A, which was made in 1985, and Zenitar-M 1,7/50, made in 1982. This image was stacked from 202 shots into one very sharp photo. Magnification is 2,5x.

Juxtaposition

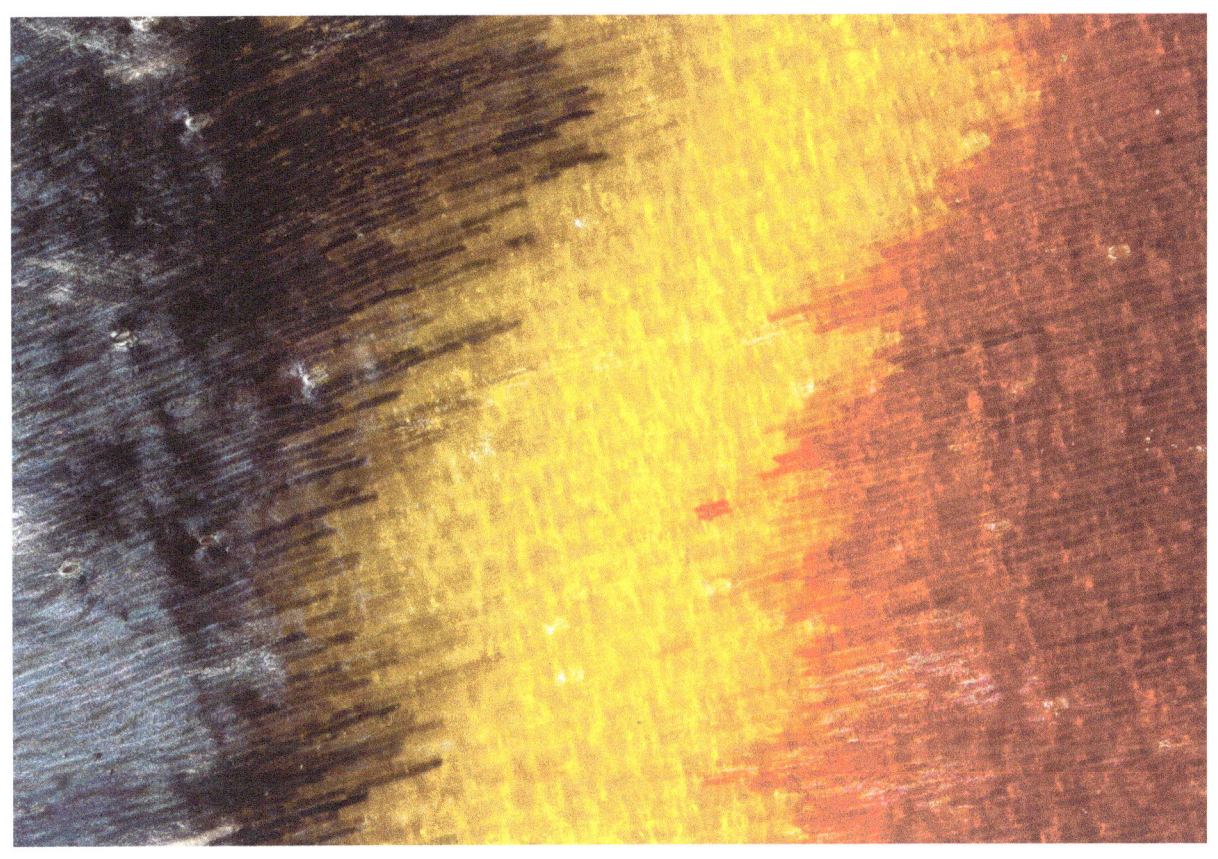

A photo of a tulip (*Tulipa*) petal taken with two lenses as one lens: MC Jupiter-37A 135mm f/3,5, which was made in 1985, and Zenitar-M 1,7/50, made in 1982. Magnification is 2.5x.

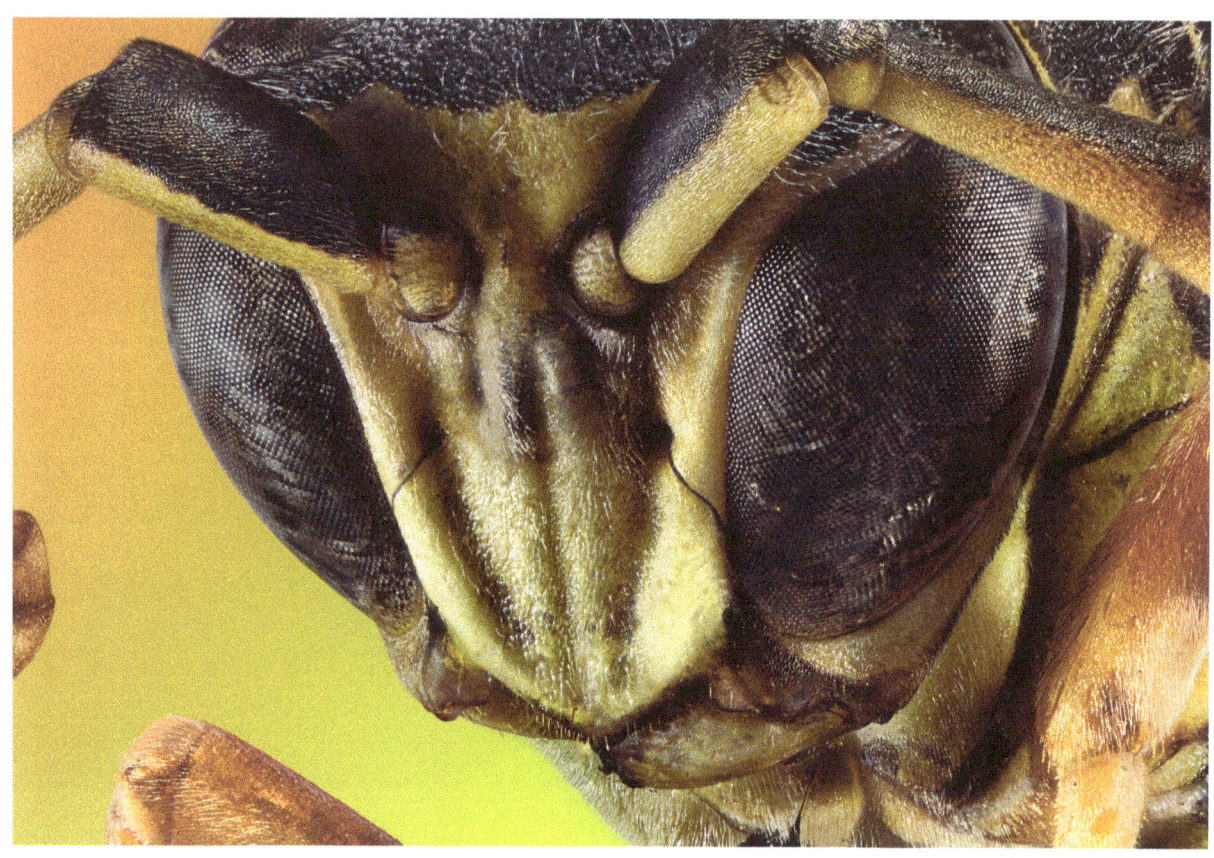

A microcosmic portrait of a paper wasp (*Polistinae*) taken with two lenses as one lens: SMC Takumar 200mm f/4, which was made in 1982, and Zenitar-M 1,7/50, made in 1982. This image stacked from 250 photos into one very sharp image. Magnification is 6x.

Juxtaposition

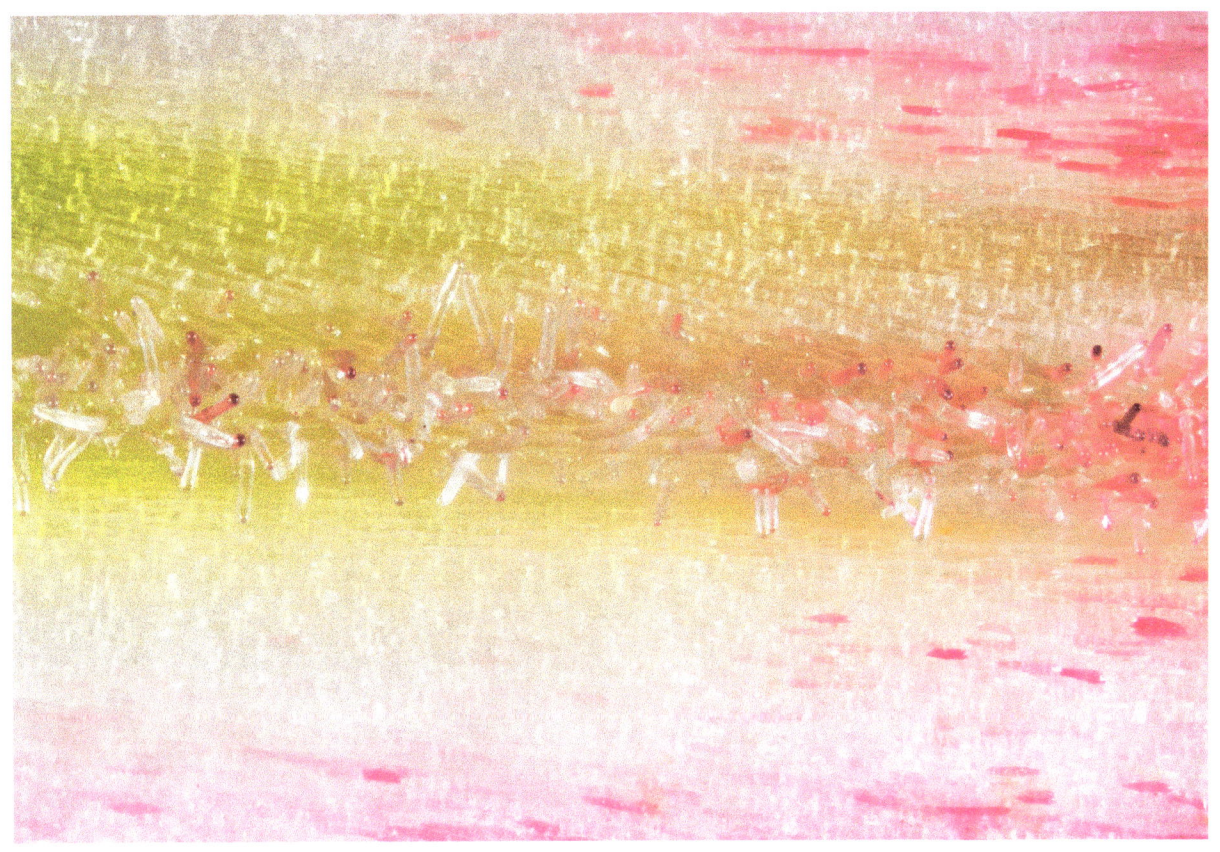

A micro photo of a lily (*Lilium*) petal taken with two lenses as one lens: SMC Takumar 200mm f/4, which was made in 1982, and Zenitar-M 1,7/50, made in 1982. This image was stacked from many shots into one very sharp photo. Magnification is 4x.

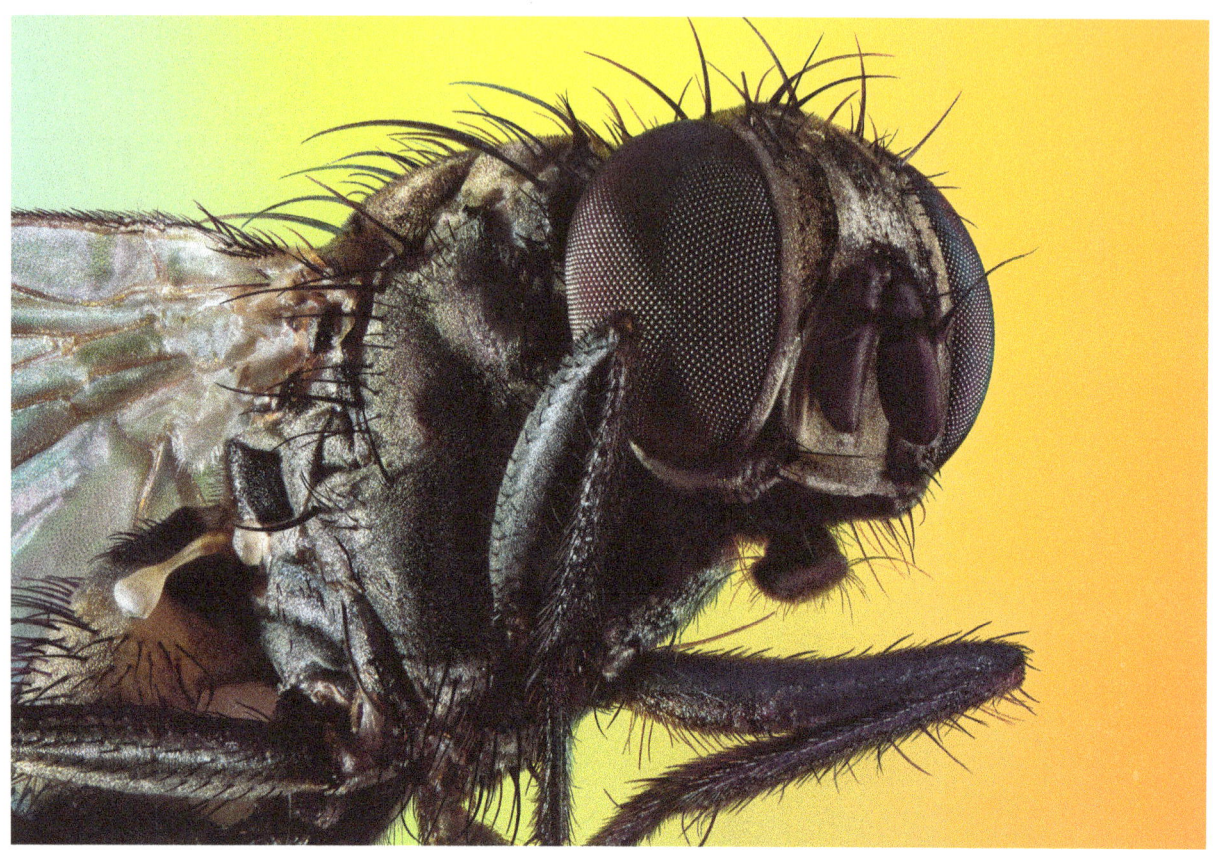

A microcosmic portrait of a housefly (*Musca domestica*) taken with two lenses as one lens: SMC Takumar 200mm f/4, which was made in 1982, and Zenitar-M 1,7/50, made in 1982. This image was stacked 163 photos into one very sharp image. Magnification is 4x.

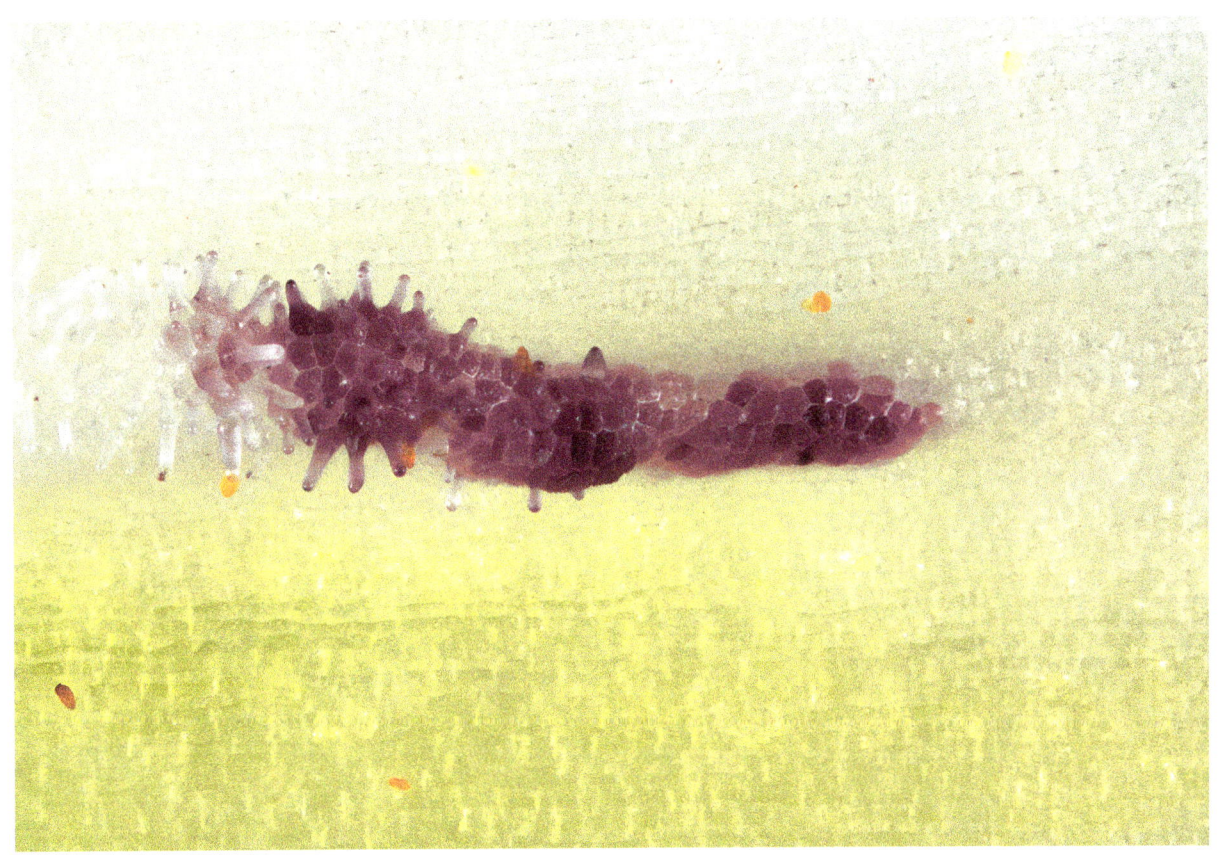

A micro photo of a lily (*Lilium*) petal taken with two lenses as one lens: SMC Takumar 200mm f/4, which was made in 1982, and Zenitar-M 1,7/50, made in 1982. This image was stacked from many shots into one very sharp photo. Magnification is 4x.

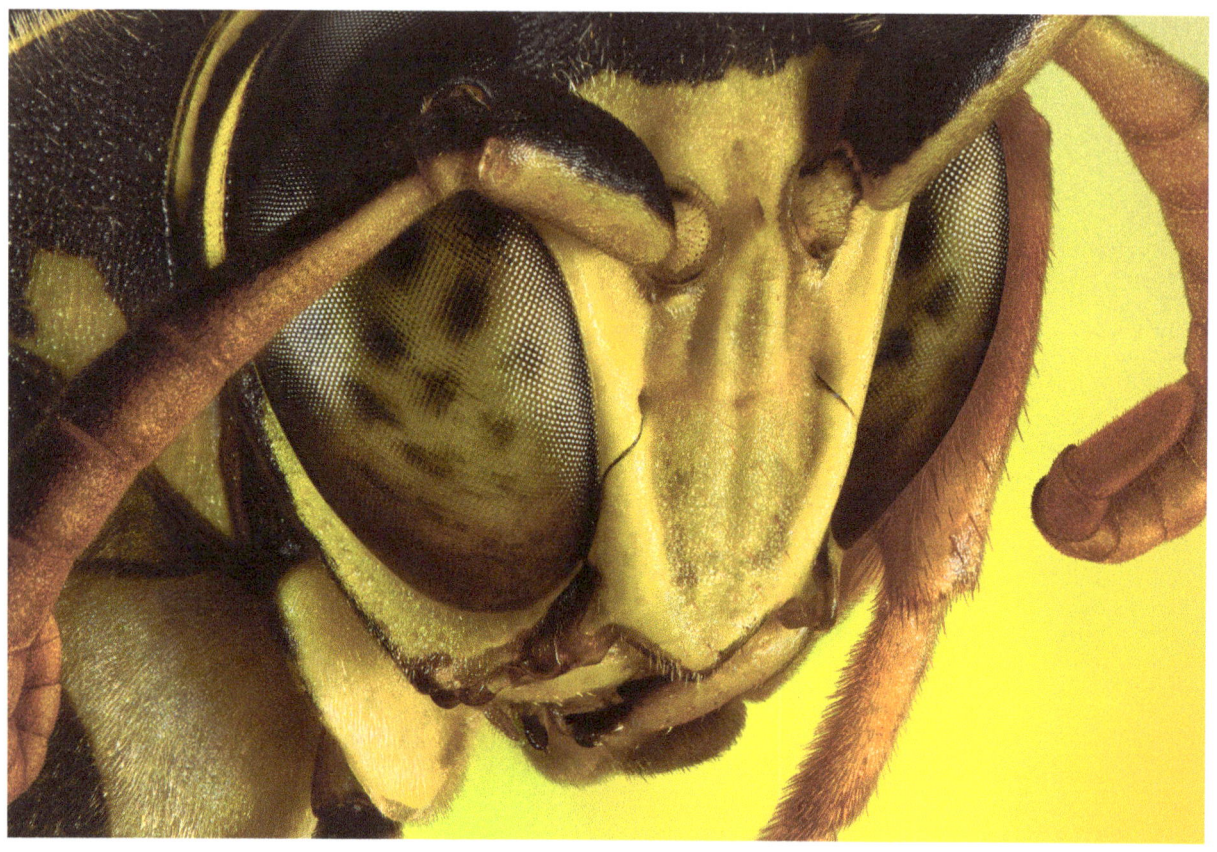

A microcosmic portrait of a paper wasp (*Polistinae*) taken with two lenses as one lens: SMC Takumar 200mm f/4, which was made in 1982, and Zenitar-M 1,7/50, made in 1982. This image was stacked from 225 photos into one very sharp image. Magnification is 6x.

Juxtaposition

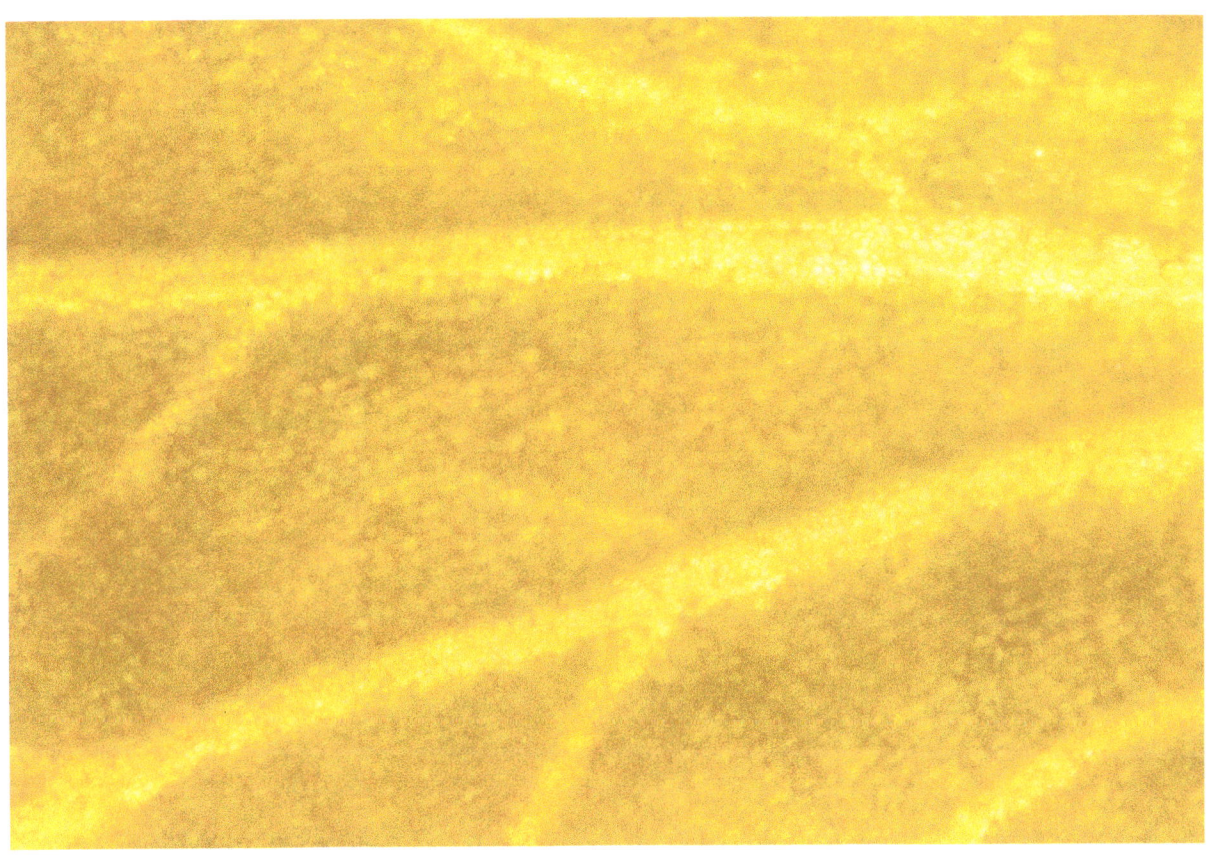

A micro photo of a rose (*Rosa*) petal taken with two lenses SMC Takumar 200mm f/4, which was made in 1982, and Zenitar-M 1,7/50, made in 1982, as one lens, stacked from many shots into one very sharp photo. Magnification is 4x.

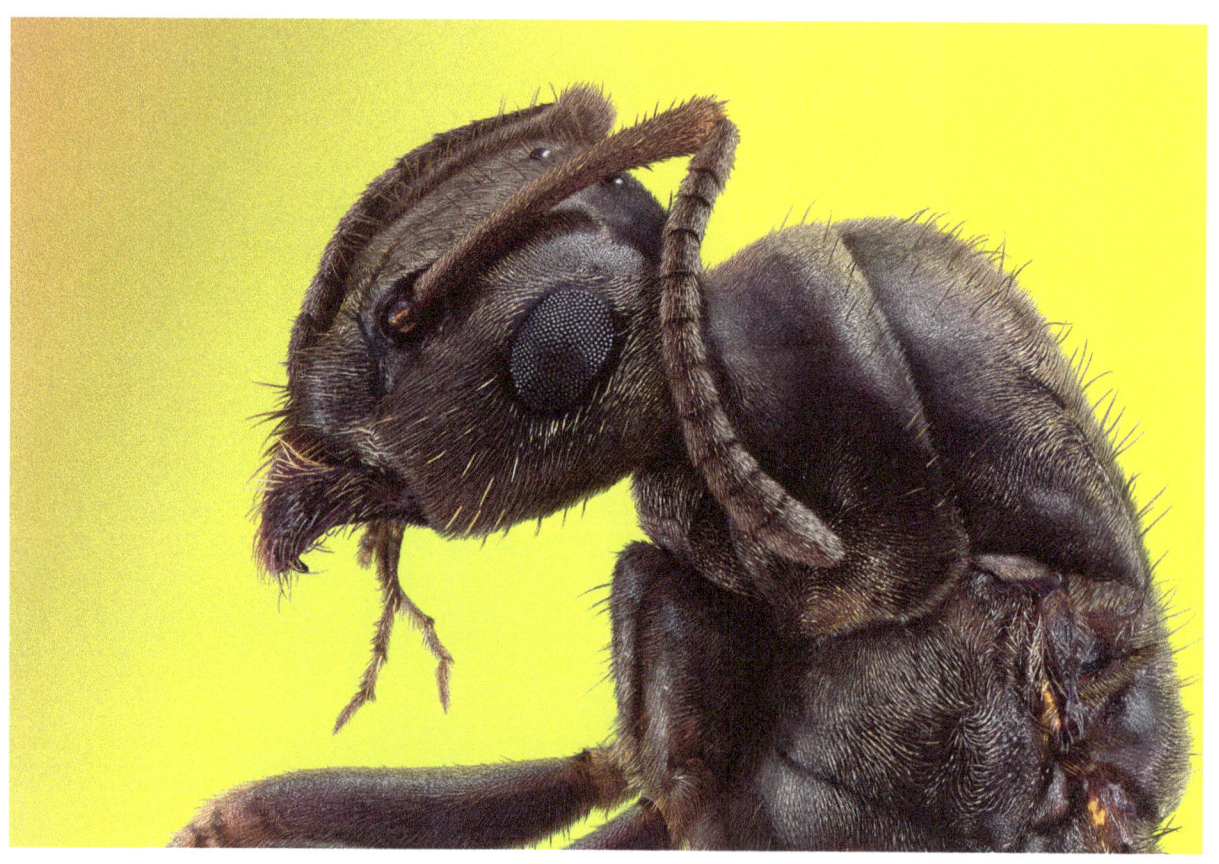

A microcosmic portrait of an ant (*Formicidae*) taken with two lenses as one lens: SMC Takumar 200mm f/4, which was made in 1982, and Zenitar-M 1,7/50, made in 1982. This image was stacked from many shots into one very sharp photo. Magnification is 4x.

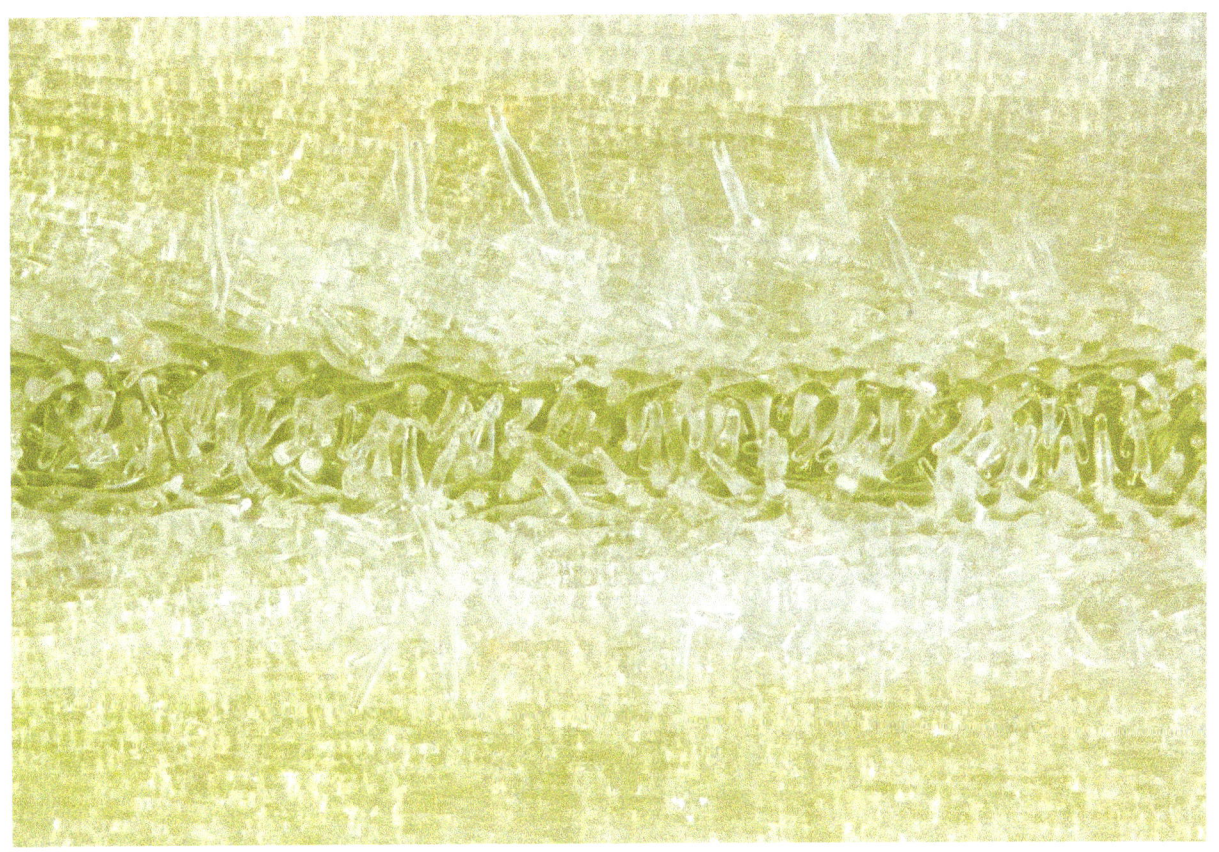

A micro photo of a lily (*Lilium*) petal taken with two lenses as one lens: SMC Takumar 200mm f/4, which was made in 1982, and Zenitar-M 1,7/50, made in 1982. This image was stacked from many shots into one very sharp photo. Magnification is 4x.

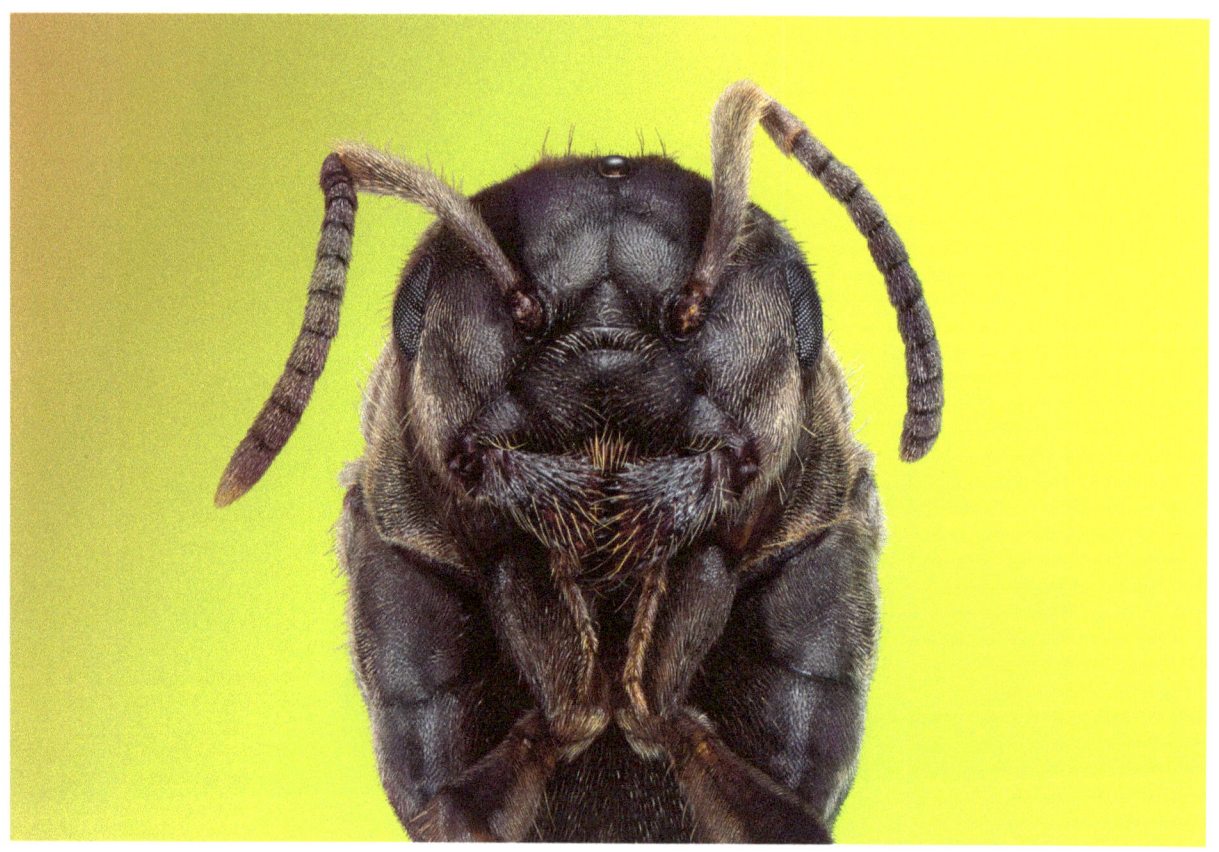

A microcosmic portrait of an ant (*Formicidae*) taken with two lenses as one lens: SMC Takumar 200mm f/4, which was made in 1982, and Zenitar-M 1,7/50, made in 1982. This image was stacked from many shots into one very sharp photo. Magnification is 4x.

Juxtaposition

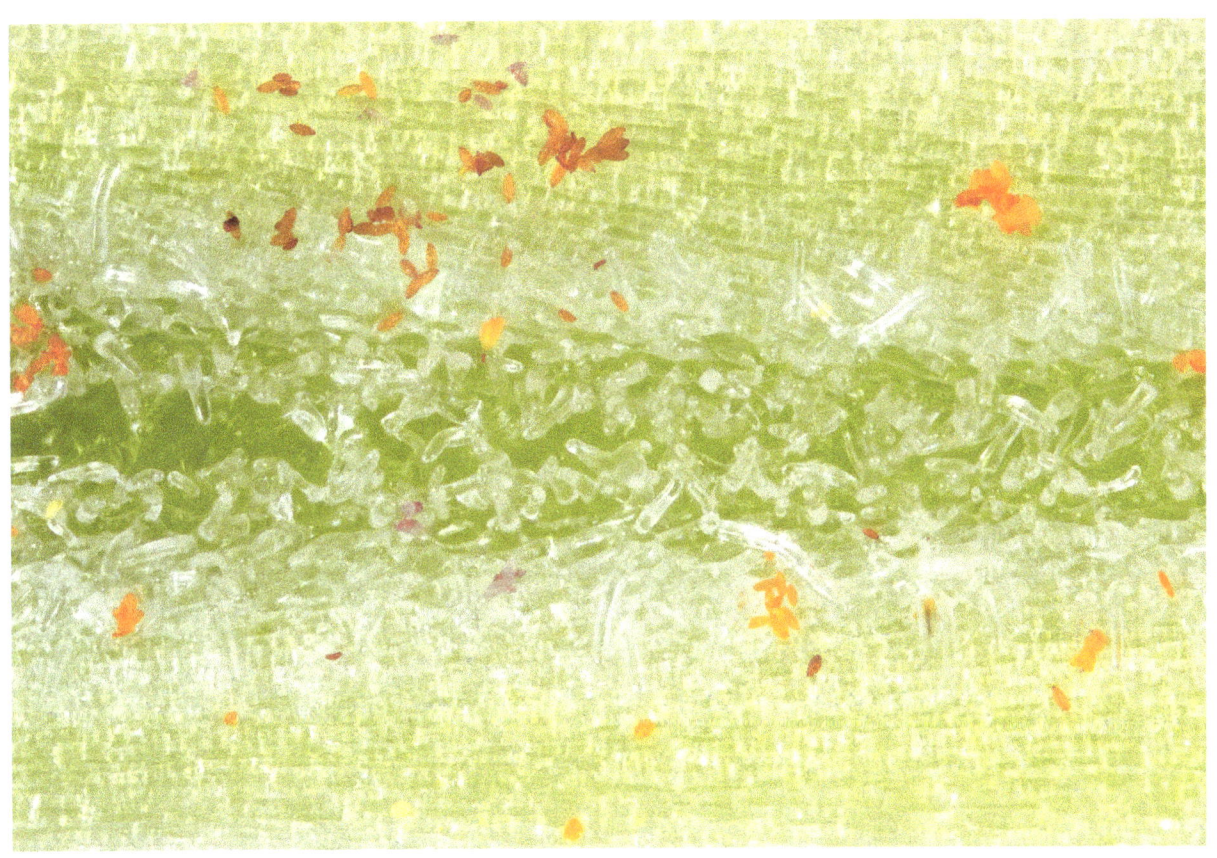

A micro photo of a lily (*Lilium*) petal taken with two lenses as one lens: SMC Takumar 200mm f/4, which was made in 1982, and Zenitar-M 1,7/50, made in 1982. This image was stacked from many shots into one very sharp photo. Magnification is 4x.

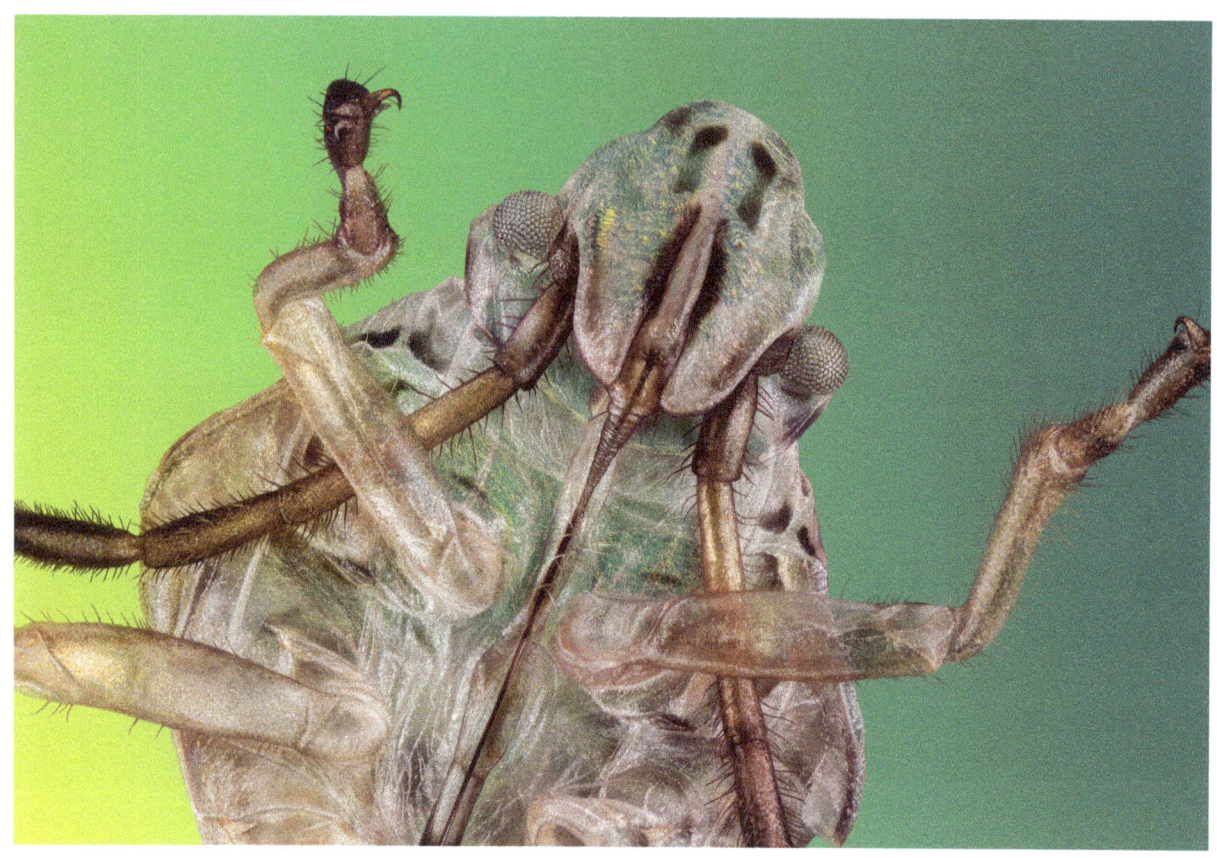

A microcosmic portrait of an exoskeleton of a shield bug or a stink bug (*Pentatomidae*) larva taken with two lenses as one lens: SMC Takumar 200mm f/4, which was made in 1982, and Zenitar-M 1,7/50, made in 1982. This image was stacked from many shots into one very sharp photo. Magnification is 4x.

Juxtaposition

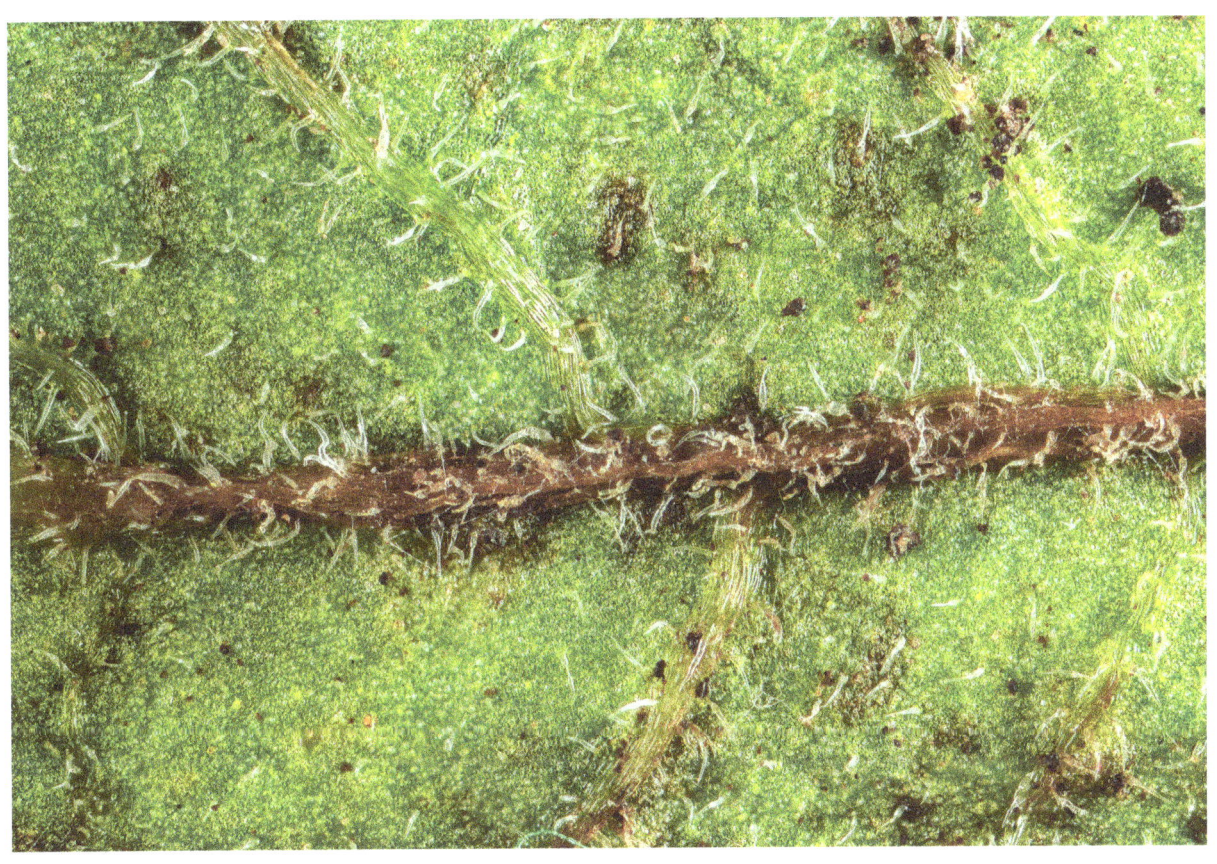

A micro photo of a potato plant (*Solanum*) leaf taken with two lenses as one lens: SMC Takumar 200mm f/4, which was made in 1982, and Zenitar-M 1,7/50, made in 1982. This image was stacked from many shots into one very sharp photo. Magnification is 4x.

Irina Petrova Adamatzky

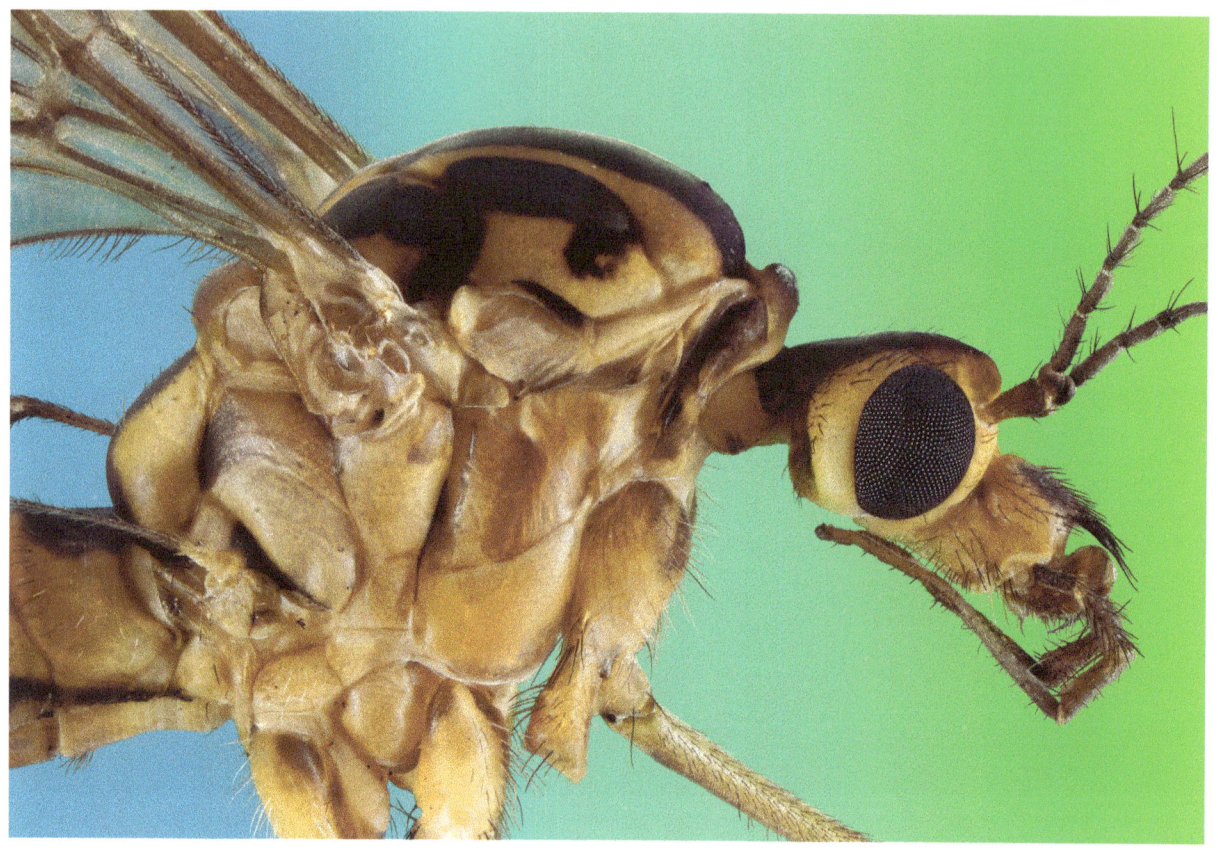

A microcosmic portrait of a crane fly (*Tipulidae*) taken with two lenses as one lens: MC Jupiter-37A, which was made in 1985, and Zenitar-M 1,7/50, made in 1982. This image was stacked from 83 shots into one very sharp photo. Magnification is 2,5x.

Juxtaposition

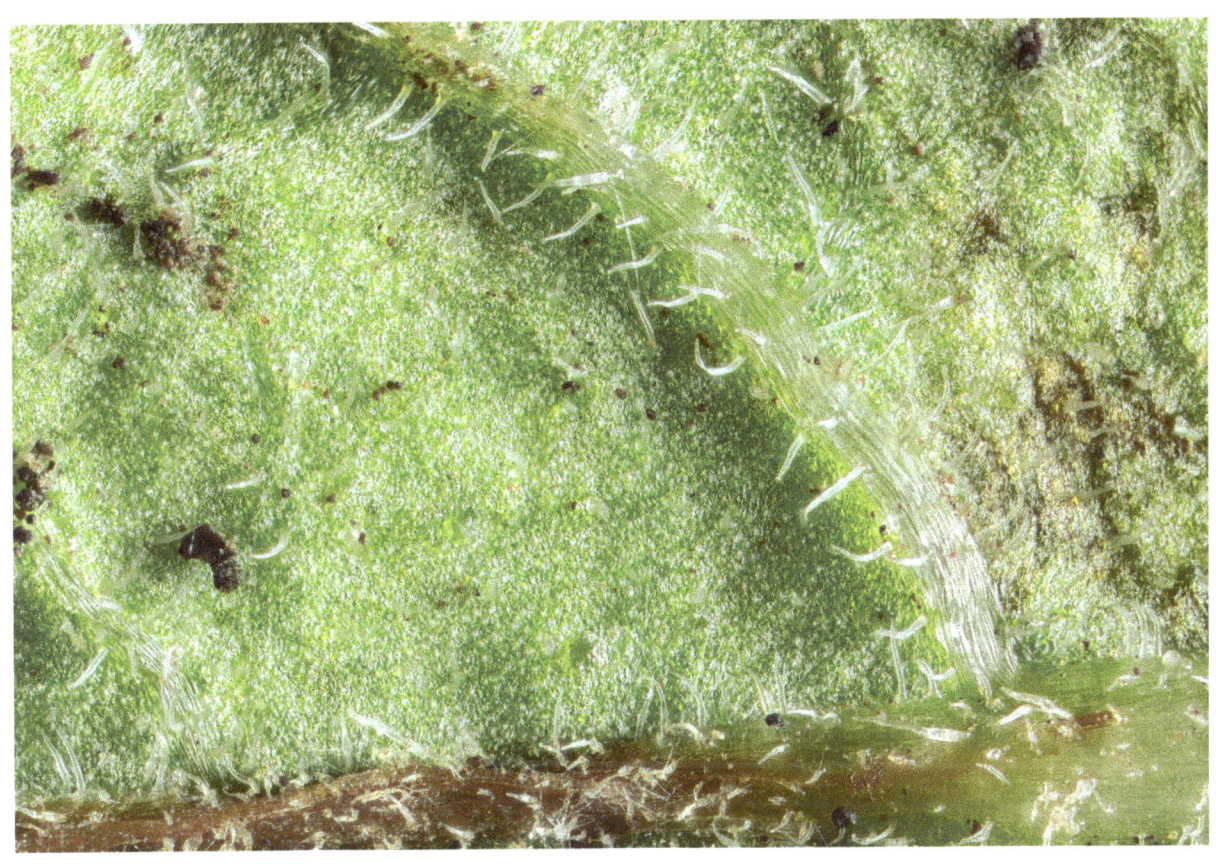

A micro photo of a potato plant (*Solanum*) leaf taken with two lenses as one lens: SMC Takumar 200mm f/4, which was made in 1982, and Zenitar-M 1,7/50, made in 1982. This image was stacked from many shots into one very sharp photo. Magnification is 4x.

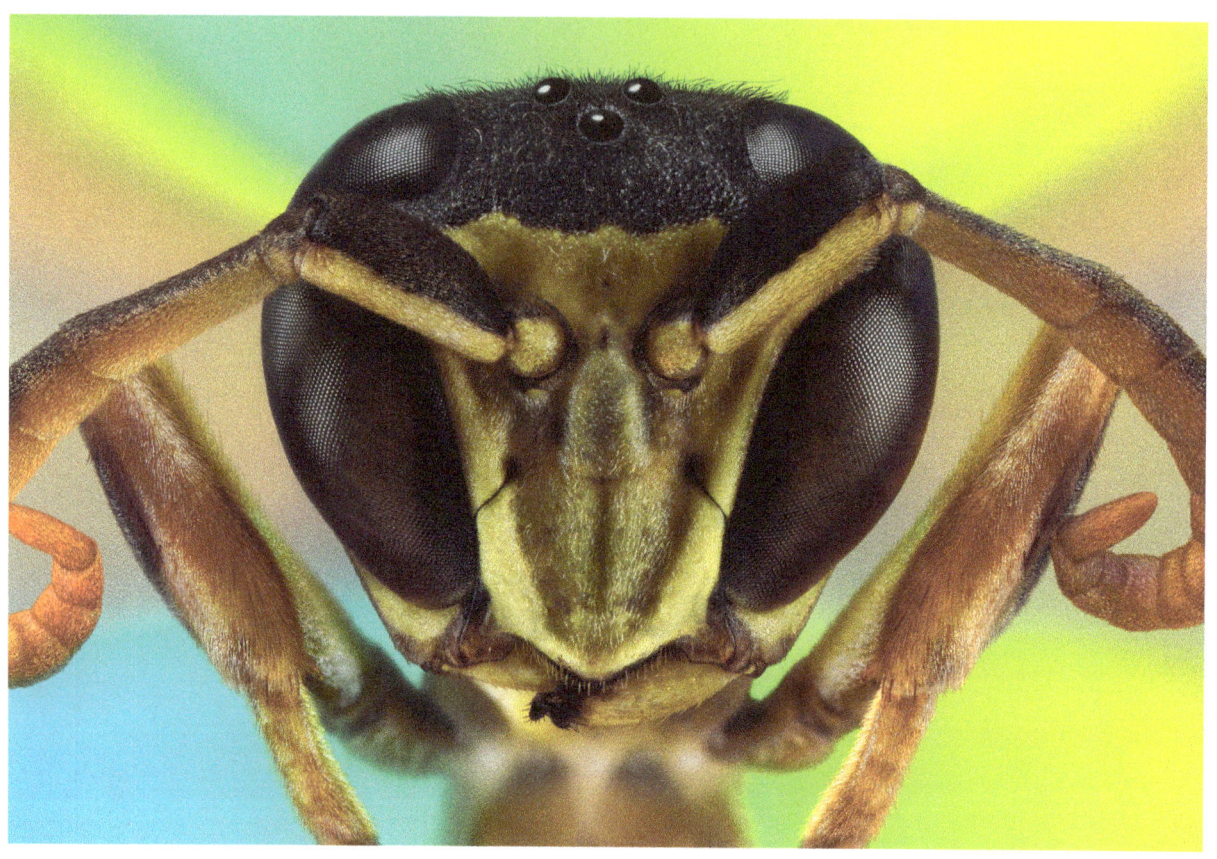

A microcosmic portrait of a wasp (*Polistinae*) taken with two lenses as one lens: MC Jupiter-37A, which was made in 1985, and Zenitar-M 1,7/50, made in 1982. This image was stacked from 70 into one very sharp photo. Magnification is 2,5x.

Juxtaposition

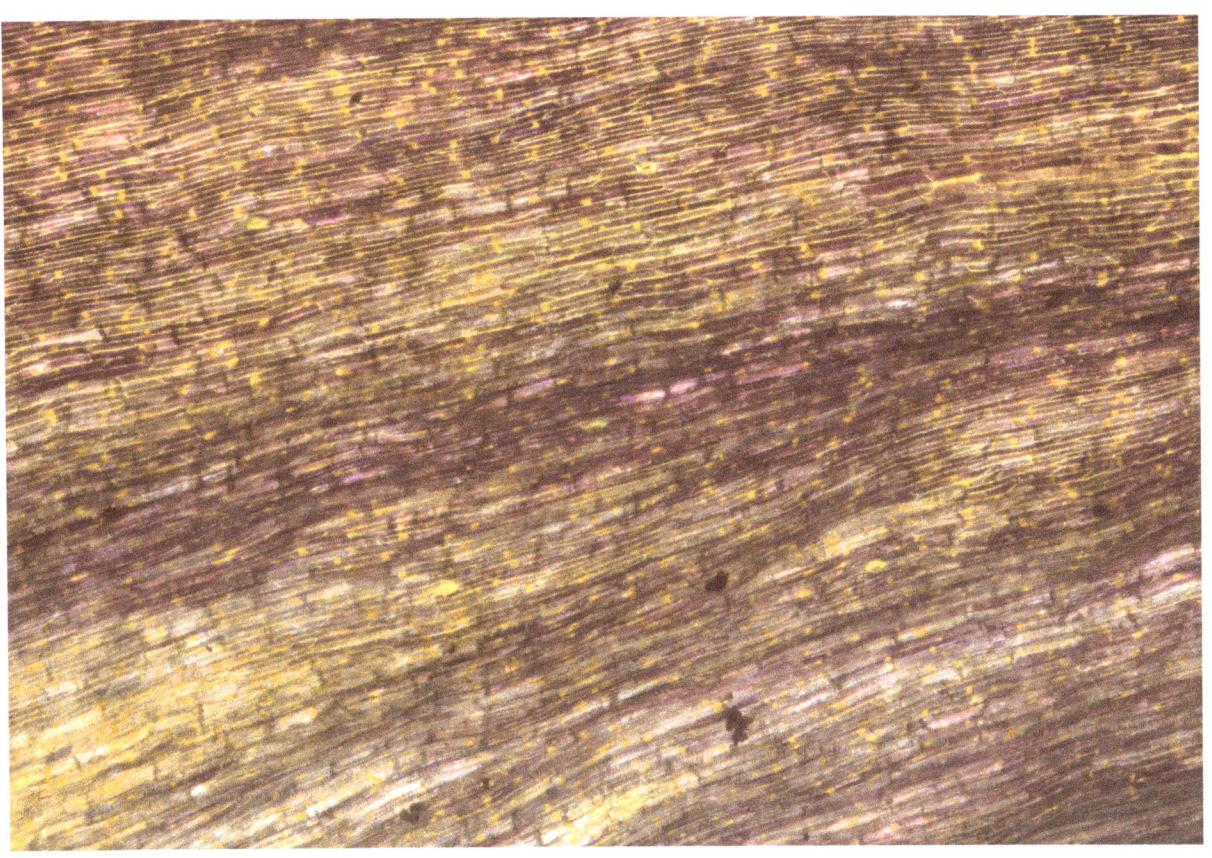

A photo of a tulip (*Tulipa*) petal taken with two lenses as one lens: MC Jupiter-37A 135mm f/3,5, which was made in 1985, and Zenitar-M 1,7/50, made in 1982. Magnification is 2.5x.

Irina Petrova Adamatzky

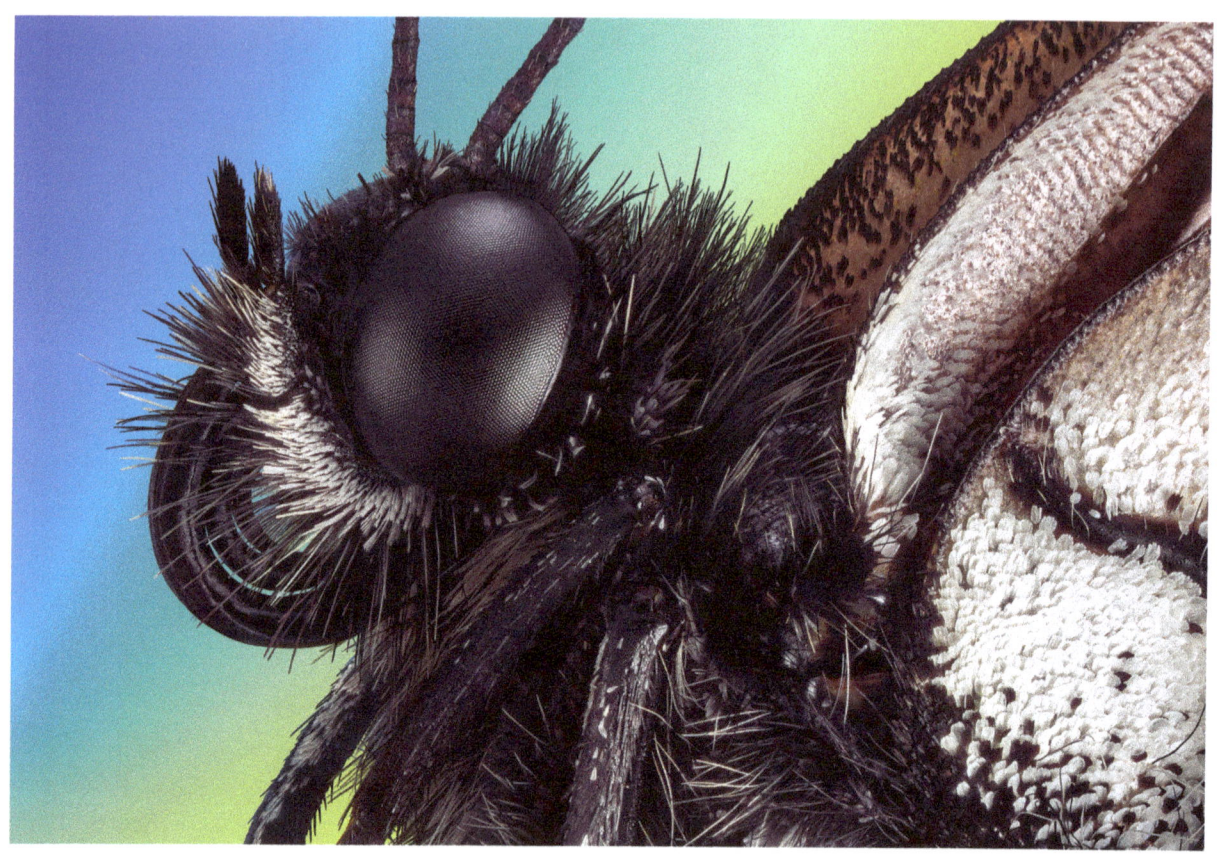

A microcosmic portrait of a butterfly (*Aporia crataegi*) taken with two lenses as one lens: MC Jupiter-37A, which was made in 1985, and Zenitar-M 1,7/50, made in 1982. This image was stacked from 112 shots into one very sharp photo. Magnification is 2,5x.

Juxtaposition

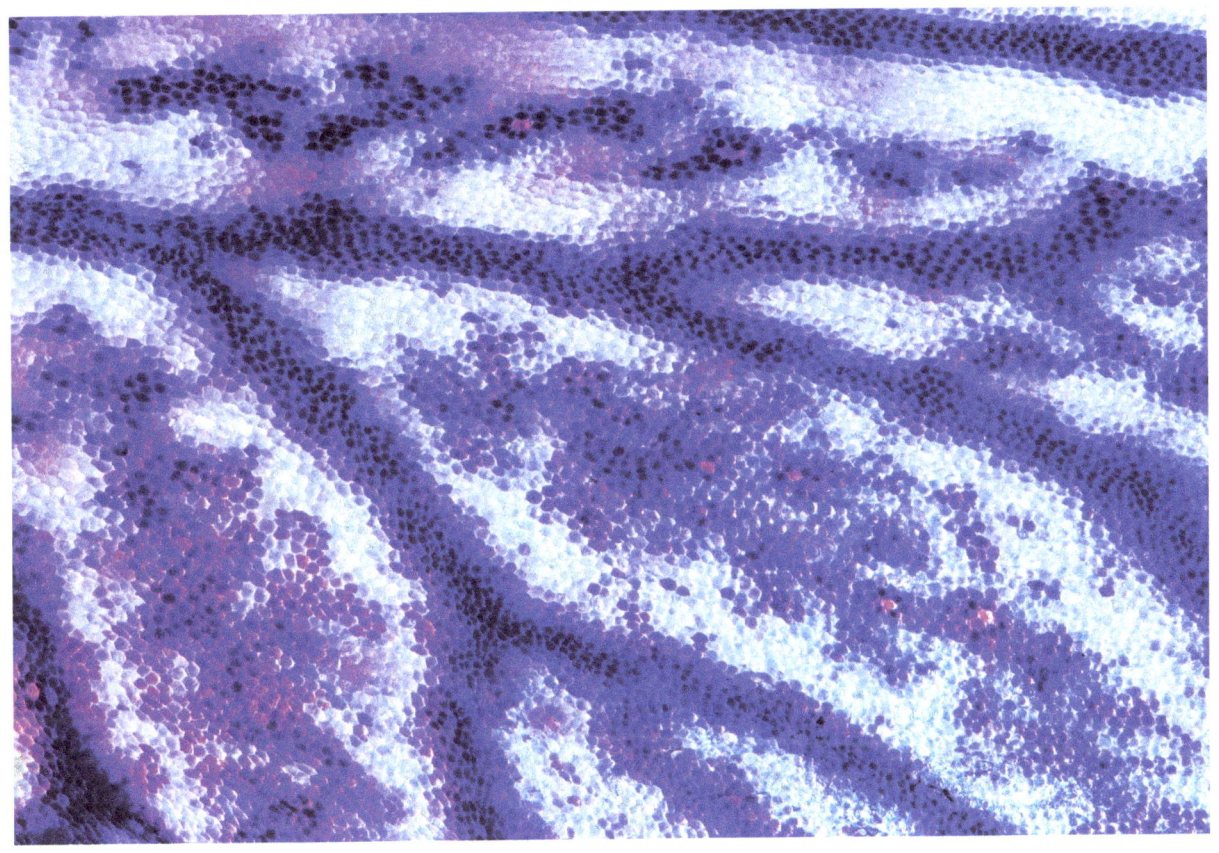

A micro photo of a wild iris petal (*Iris*) taken with two lenses as one lens: SMC Takumar 200mm f/4, which was made in 1982, and a retro lens Zenitar-M 1,7/50, made in 1982. This image was stacked from many shots into one very sharp photo. Magnification is 4x.

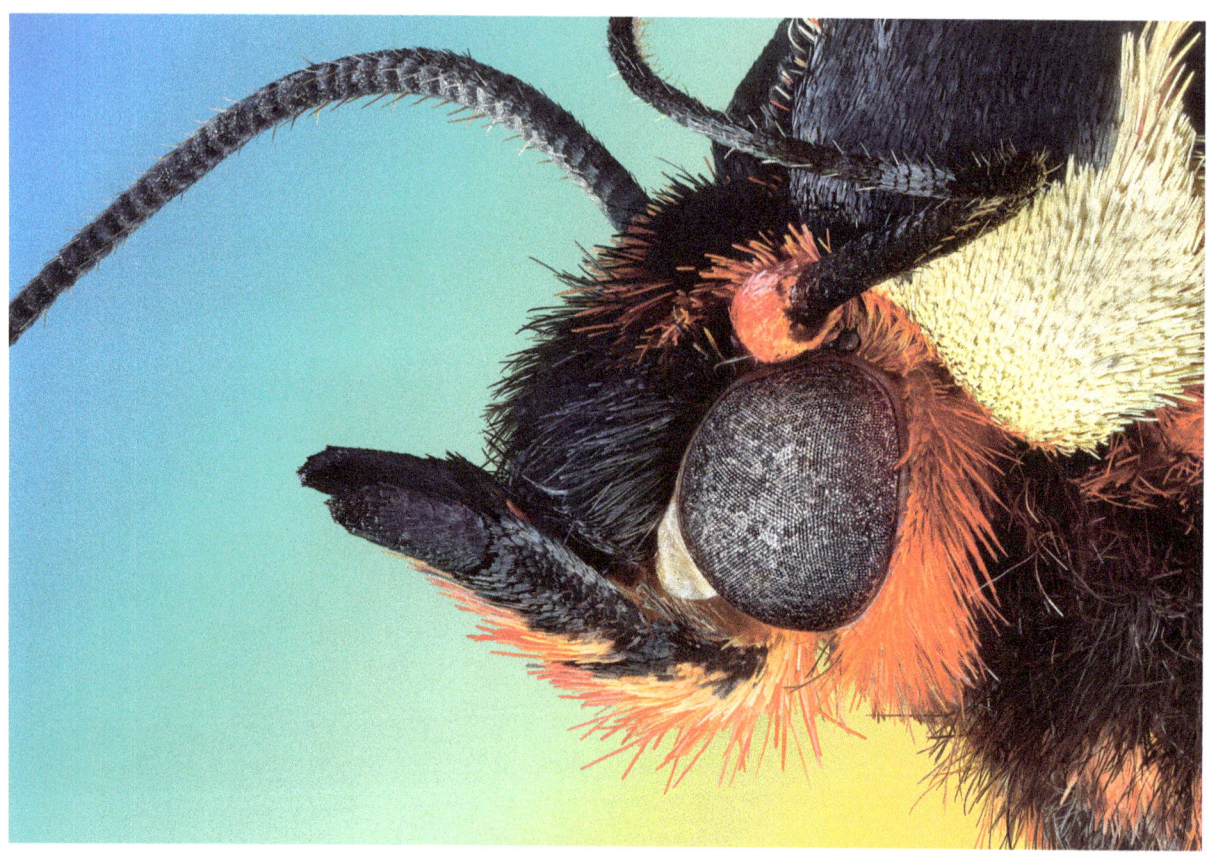

A microcosmic portrait of a garden tiger moth (*Arctia caja*) taken with two lenses as one lens: MC Jupiter-37A, which was made in 1985, and Zenitar-M 1,7/50, made in 1982. This image was stacked from 179 shots into one very sharp photo. Magnification is 2,5x.

Juxtaposition

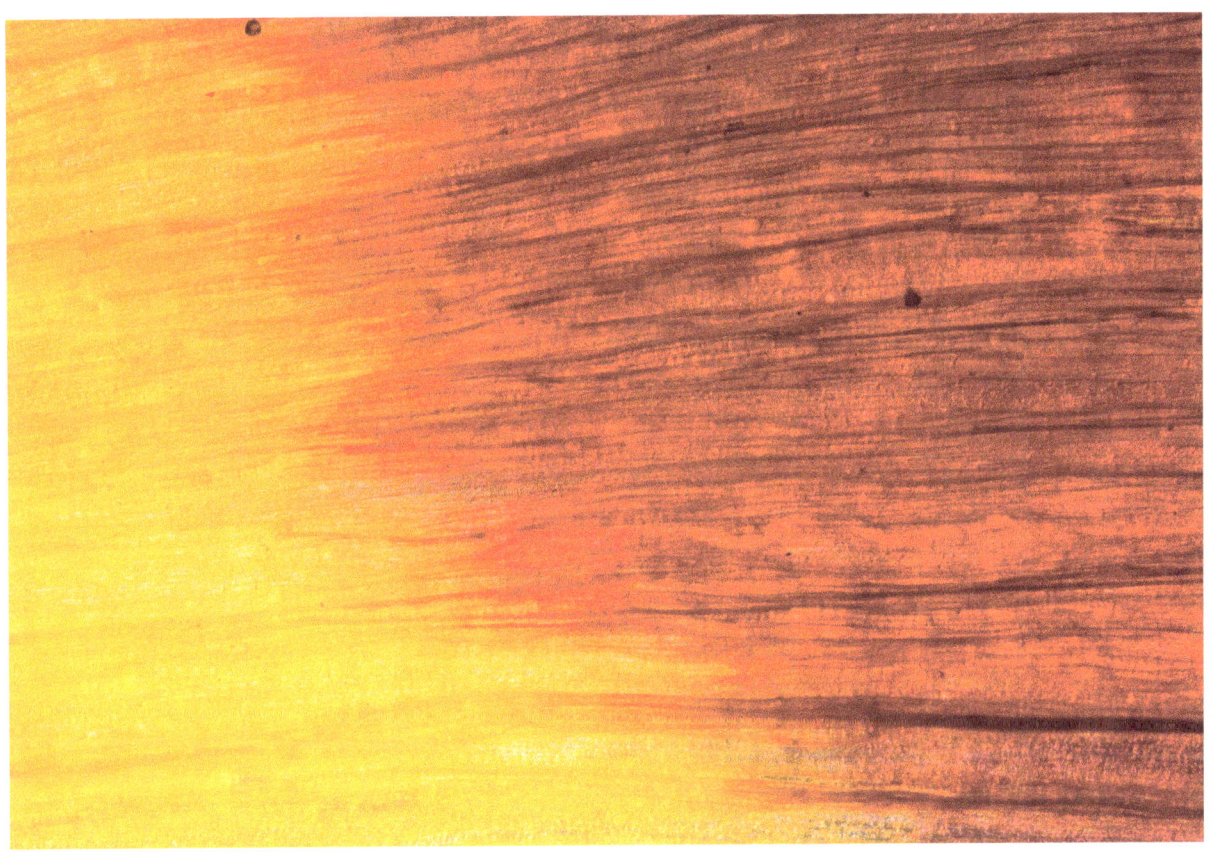

A photo of a tulip (*Tulipa*) petal taken with two lenses as one lens: MC Jupiter-37A 135mm f/3,5, which was made in 1985, and Zenitar-M 1,7/50, made in 1982. Magnification is 2.5x.

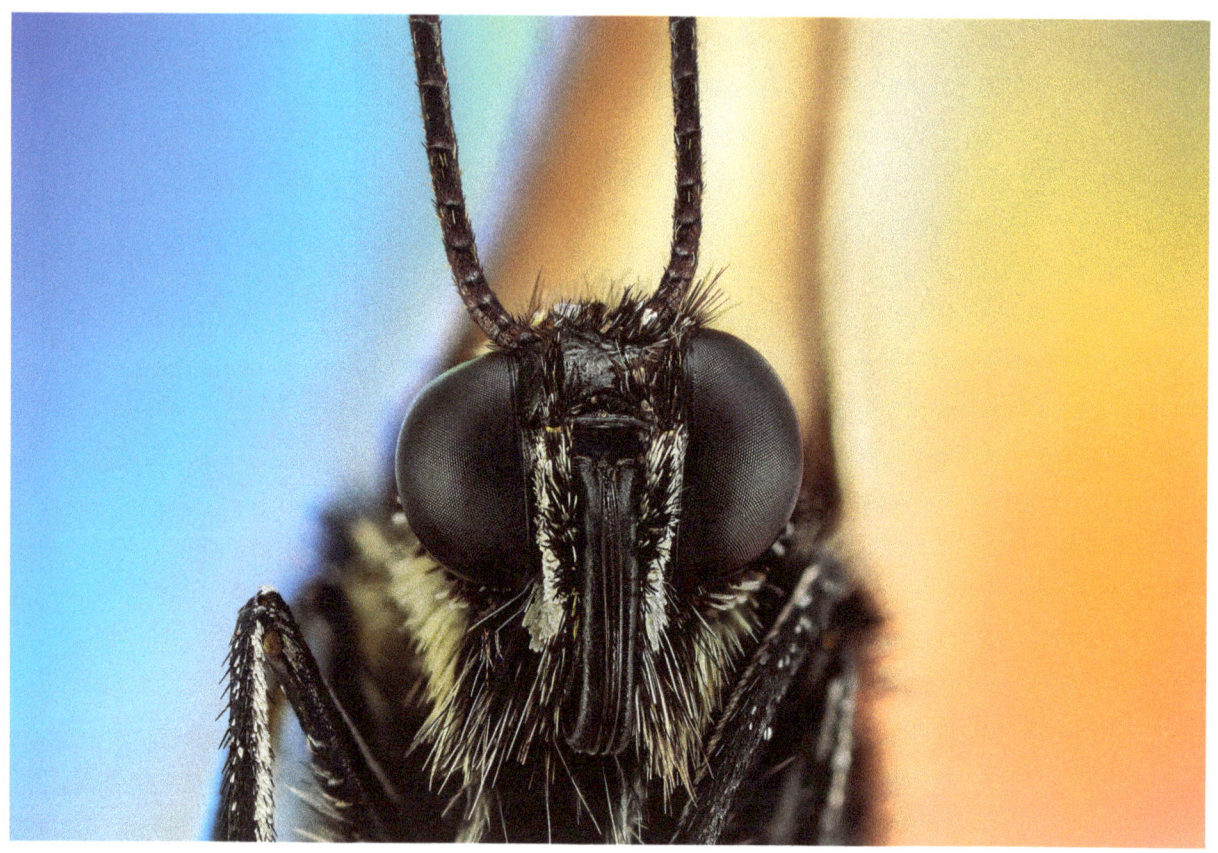

A microcosmic portrait of a butterfly (*Aporia crataegi*) taken with two lenses as one lens: MC Jupiter-37A, which was made in 1985, and Zenitar-M 1,7/50, made in 1982. This image was stacked from 93 shots into one very sharp photo. Magnification is 2,5x.

Juxtaposition

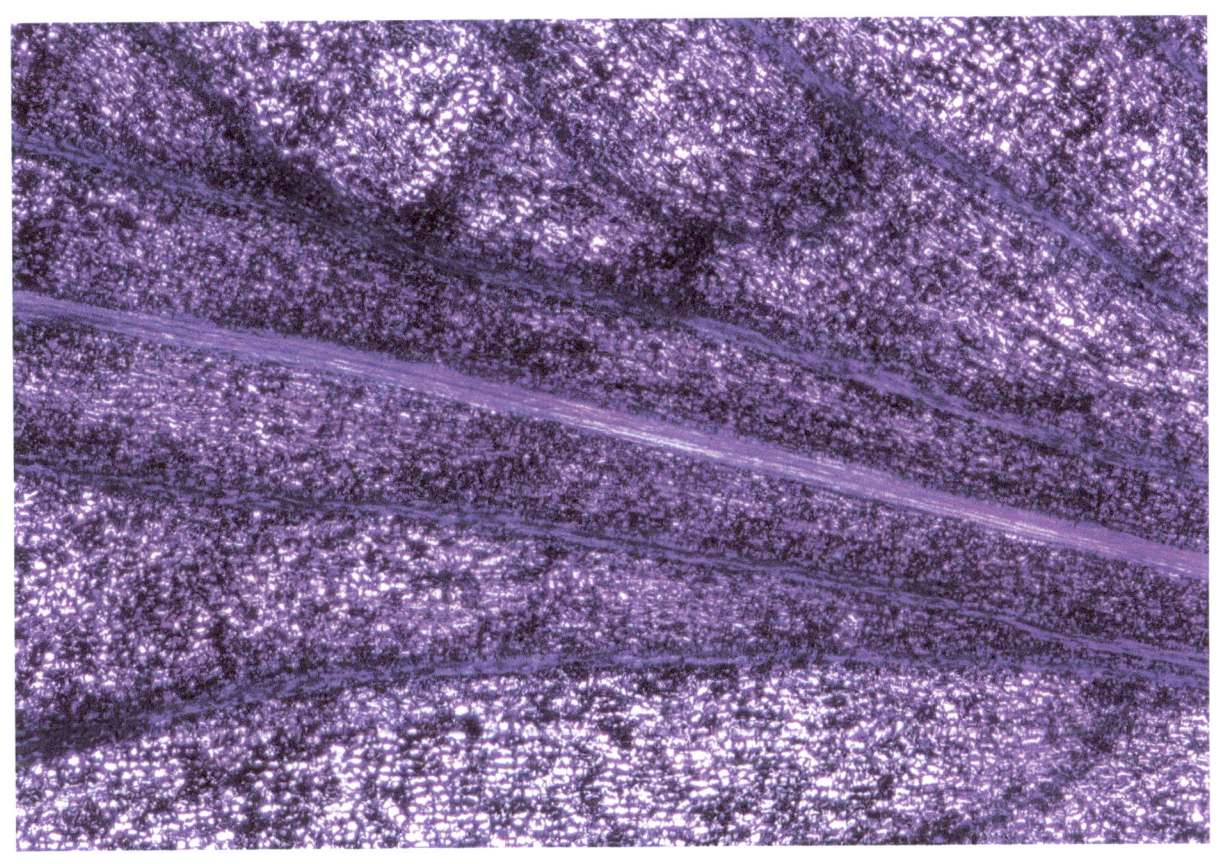

A micro photo of a wild iris petal (*Iris*) taken with two lenses as one lens: SMC Takumar 200mm f/4, which was made in 1982, and a retro lens Zenitar-M 1,7/50, made in 1982. This image was stacked from many shots into one very sharp photo. Magnification is 4x.

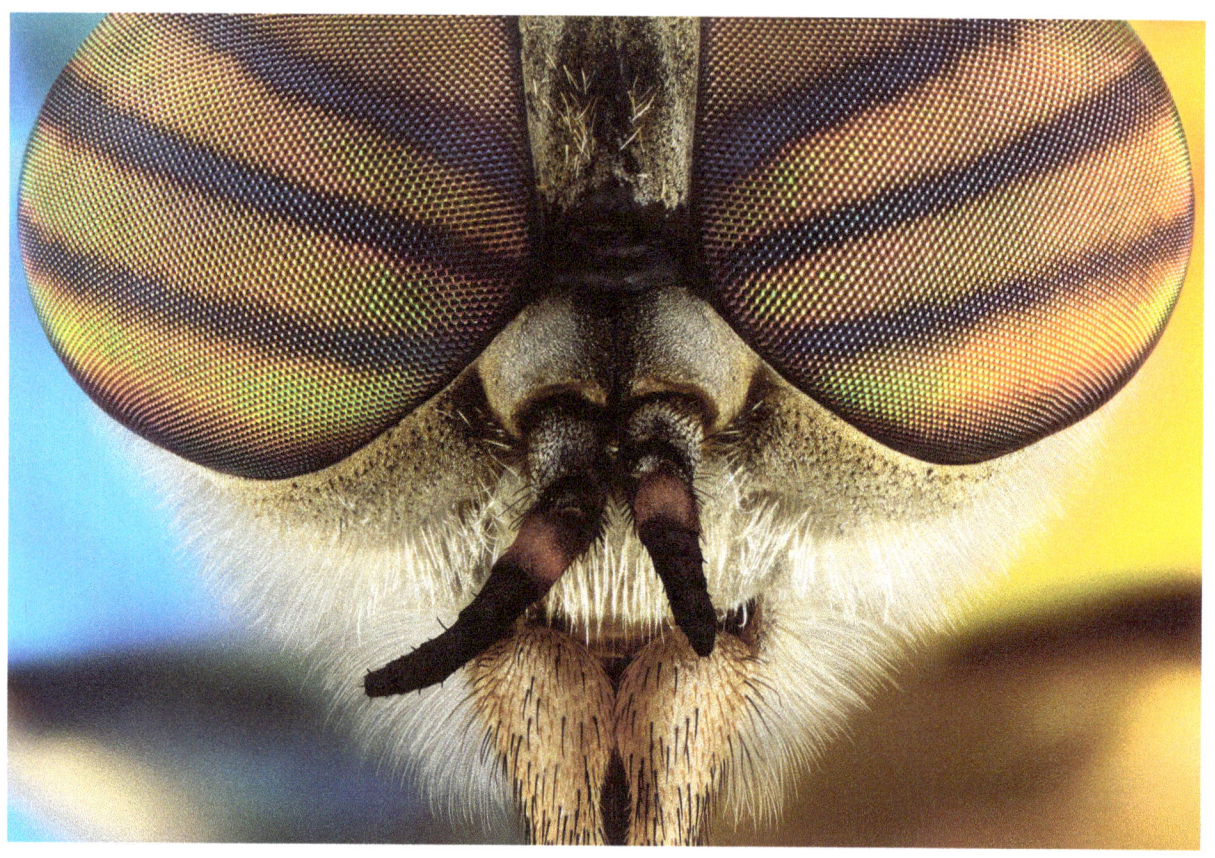

A microcosmic portrait of a horsefly or a gadfly (*Tabanidae*) taken with two lenses as one lens: SMC Takumar 200mm f/4, which was made in 1982, and Zenitar-M 1,7/50, made in 1982. This image was stacked from 150 photos into one very sharp image. Magnification is 4x.

Juxtaposition

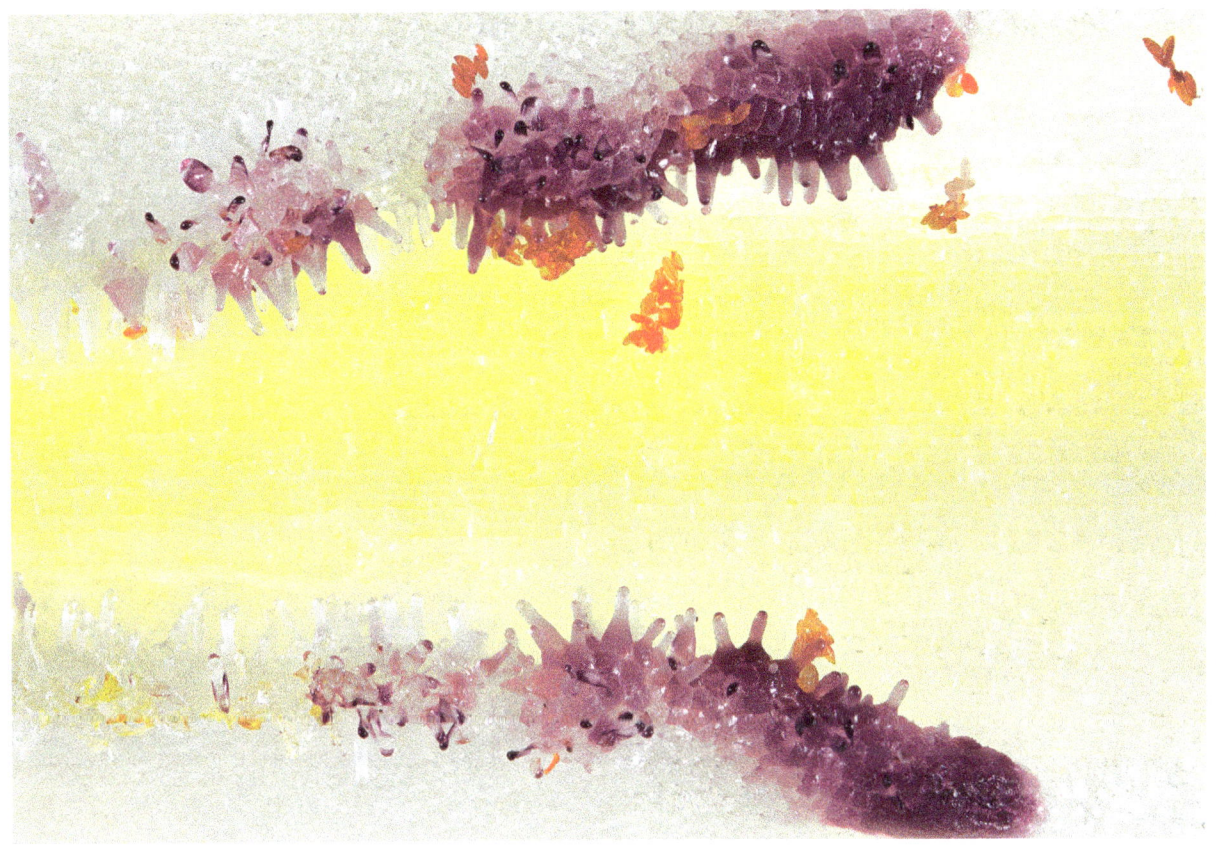

A micro photo of a lily (*Lilium*) petal taken with two lenses as one lens: SMC Takumar 200mm f/4, which was made in 1982, and Zenitar-M 1,7/50, made in 1982. This image was stacked from many shots into one very sharp photo. Magnification is 4x.

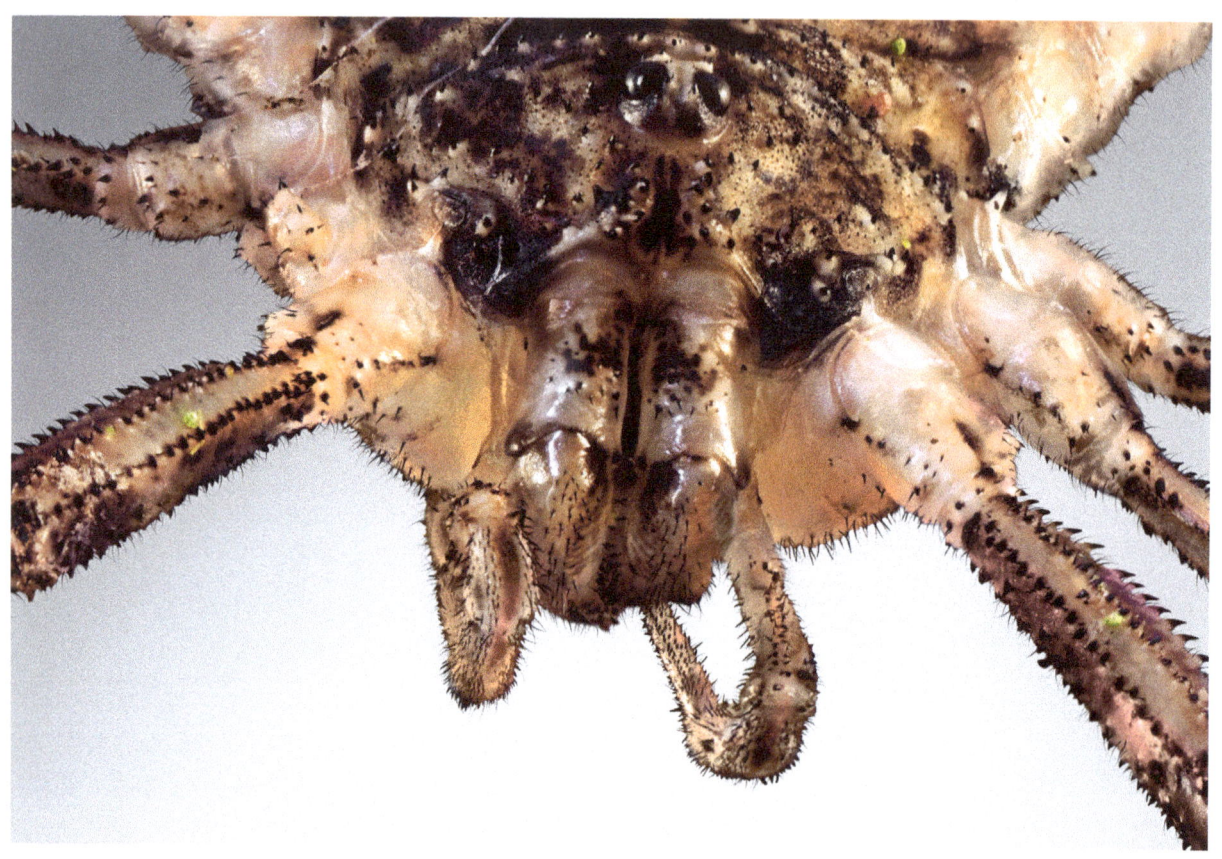

A microcosmic portrait of a harvestman (*Opiliones*), which is also known as harvester or daddy longlegs, taken with two lenses as one lens: SMC Takumar 200mm f/4, which was made in 1982, and Zenitar-M 1,7/50, made in 1982. This image was stacked from many shots into one very sharp photo. Magnification is 4x.

Juxtaposition

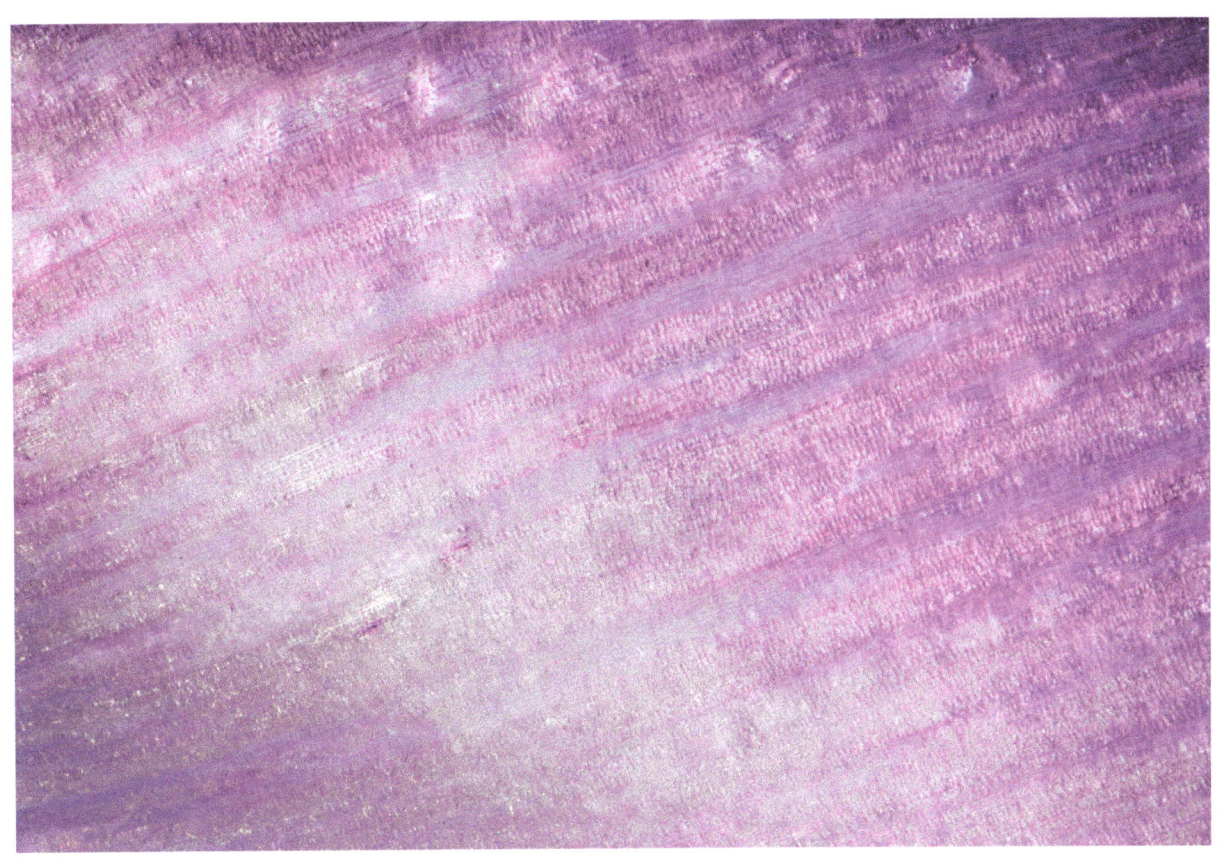

A photo of a tulip (*Tulipa*) petal taken with two lenses as one lens: MC Jupiter-37A 135mm f/3,5, which was made in 1985, and Zenitar-M 1,7/50, made in 1982. Magnification is 2.5x.

Irina Petrova Adamatzky

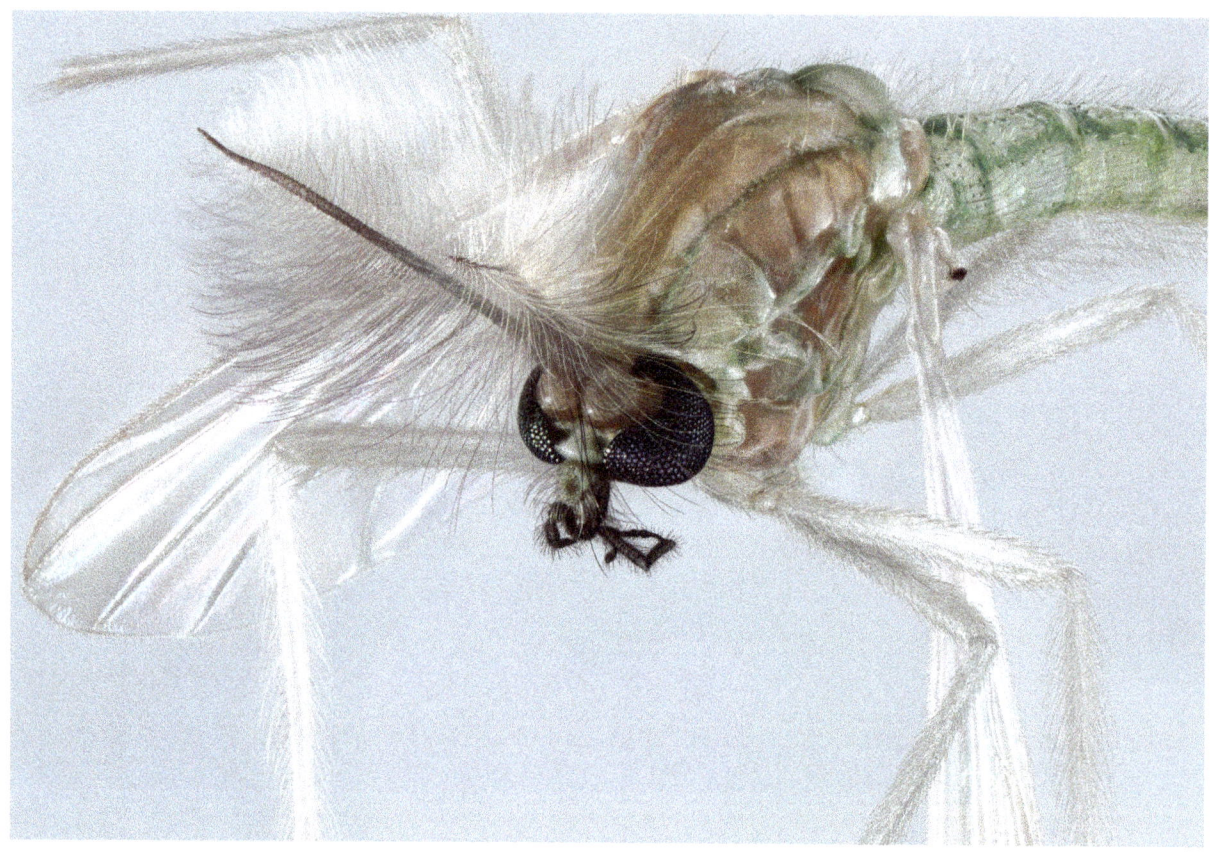

A microcosmic portrait of mosquito (*Chironomidae*), which is often called a lake fly, chironomid or a nonbiting midge, taken with two lenses as one lens: SMC Takumar 200mm f/4, which was made in 1982, and Zenitar-M 1,7/50, made in 1982. This image was stacked from many shots into one very sharp photo. Magnification is 4x.

Juxtaposition

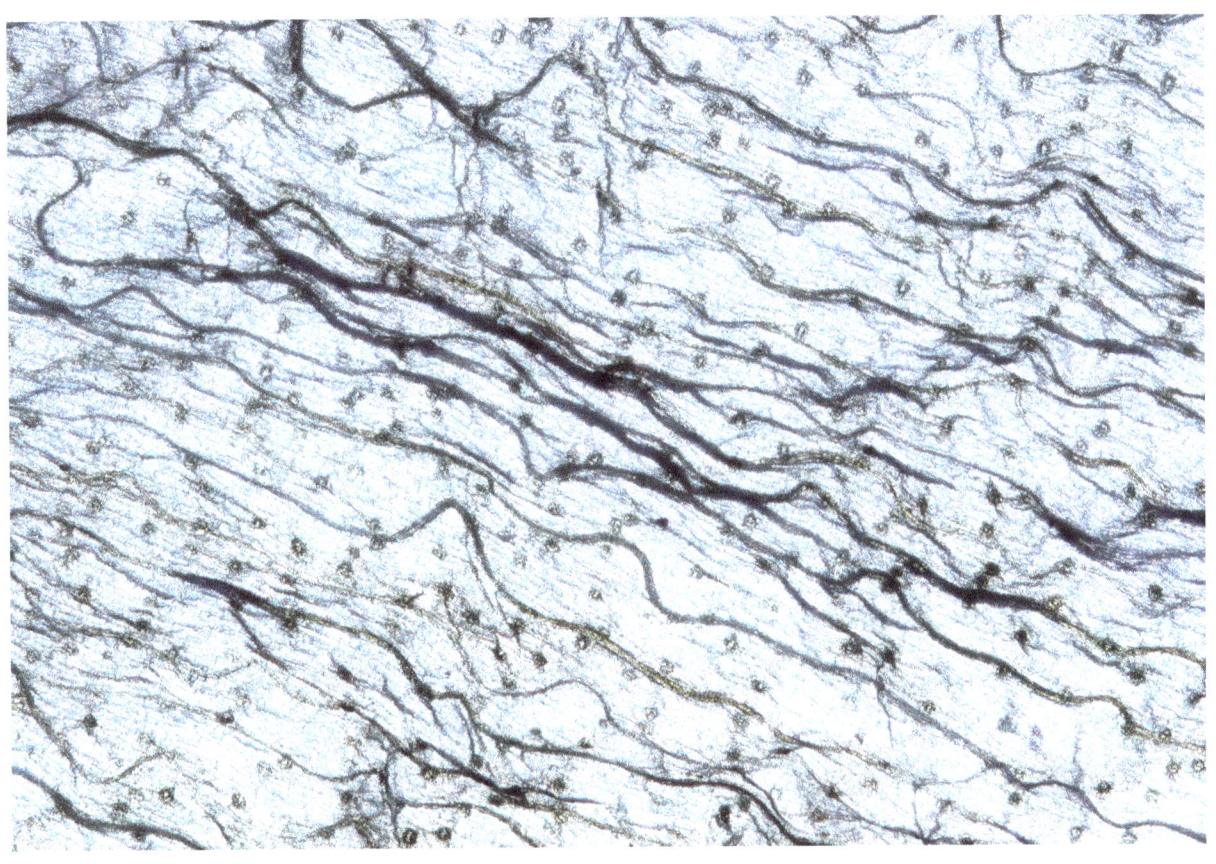

A photo of a tulip (*Tulipa*) petal taken with two lenses as one lens: MC Jupiter-37A 135mm f/3,5, which was made in 1985, and Zenitar-M 1,7/50, made in 1982. Magnification is 2.5x.

Irina Petrova Adamatzky

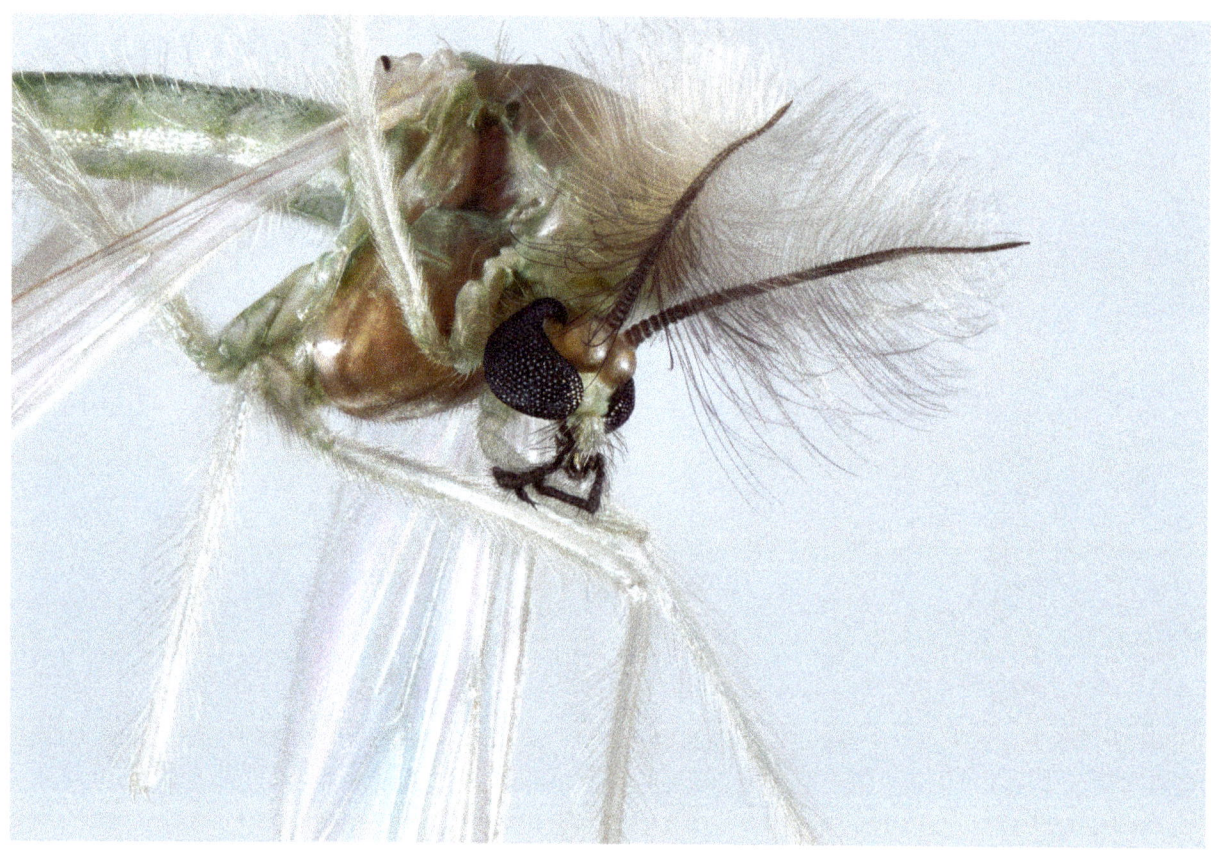

A microcosmic portrait of mosquito (*Chironomidae*), which is often called a lake fly, chironomid or a nonbiting midge, taken with two lenses as one lens: SMC Takumar 200mm f/4, which was made in 1982, and Zenitar-M 1,7/50, made in 1982. This image was stacked from many shots into one very sharp photo. Magnificafion is 4x.

Juxtaposition

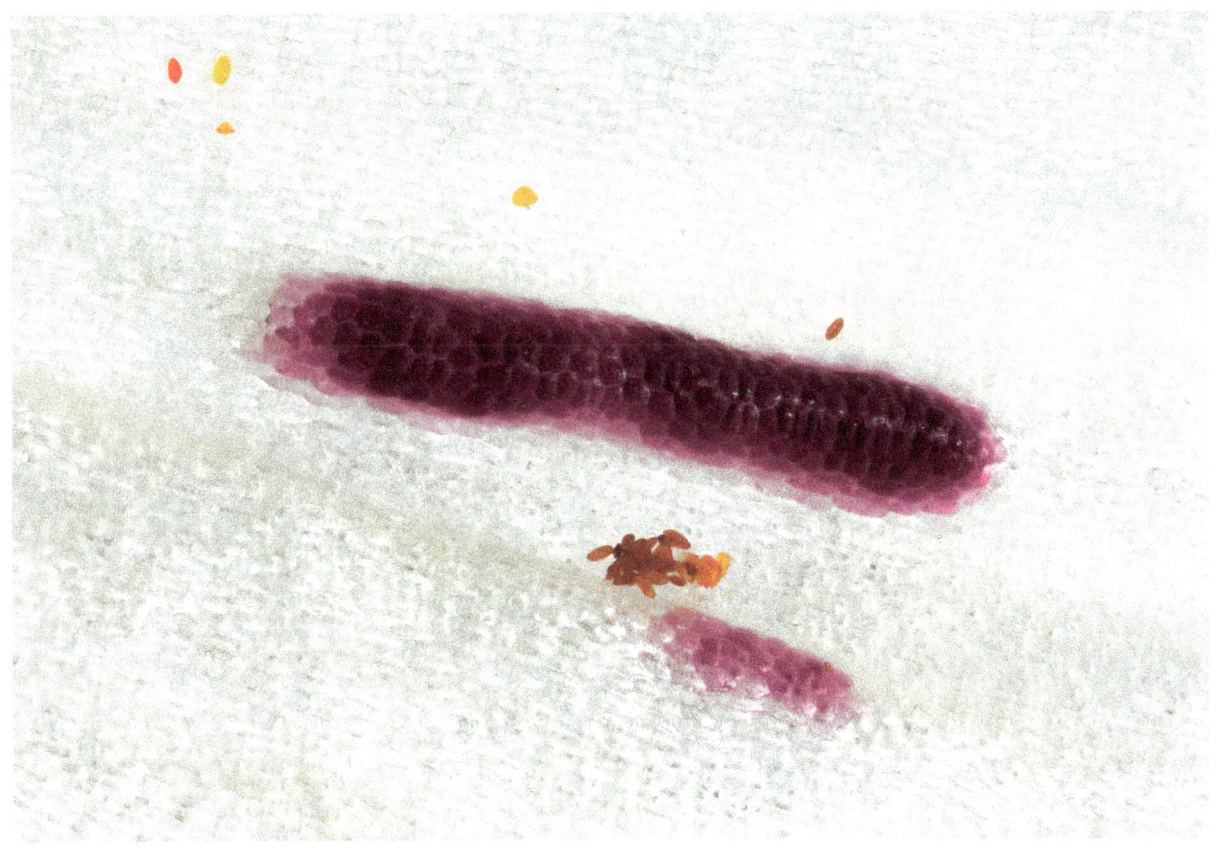

A micro photo of a lily *(Lilium)* petal taken with two lenses as one lens: SMC Takumar 200mm f/4, which was made in 1982, and Zenitar-M 1,7/50, made in 1982. This image was stacked from many shots into one very sharp photo. Magnification is 4x.

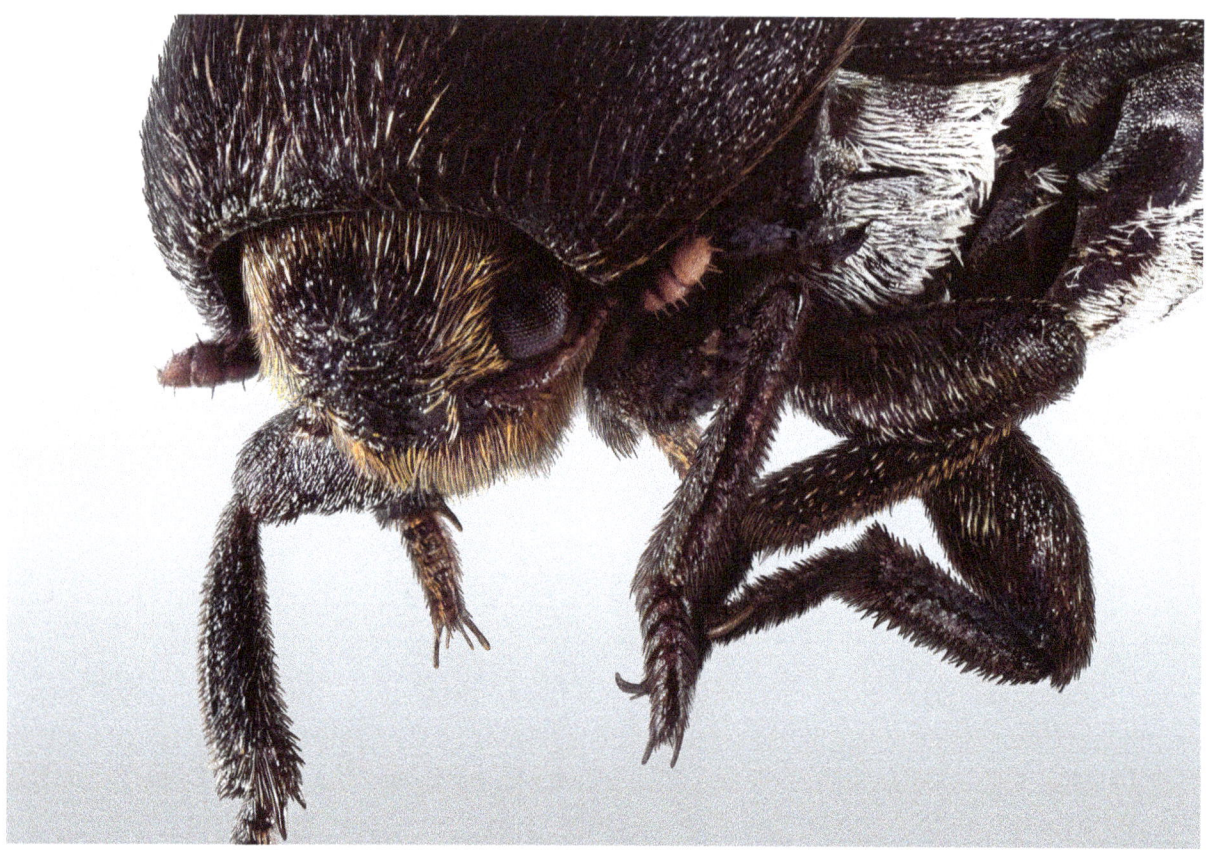

A microcosmic portrait of a drugstore beetle (*Stegobium paniceum*) taken with two lenses as one lens: SMC Takumar 200mm f/4, which was made in 1982, and Zenitar-M 1,7/50, made in 1982. This image was stacked from many shots into one very sharp photo. Magnification is 4x.

Juxtaposition

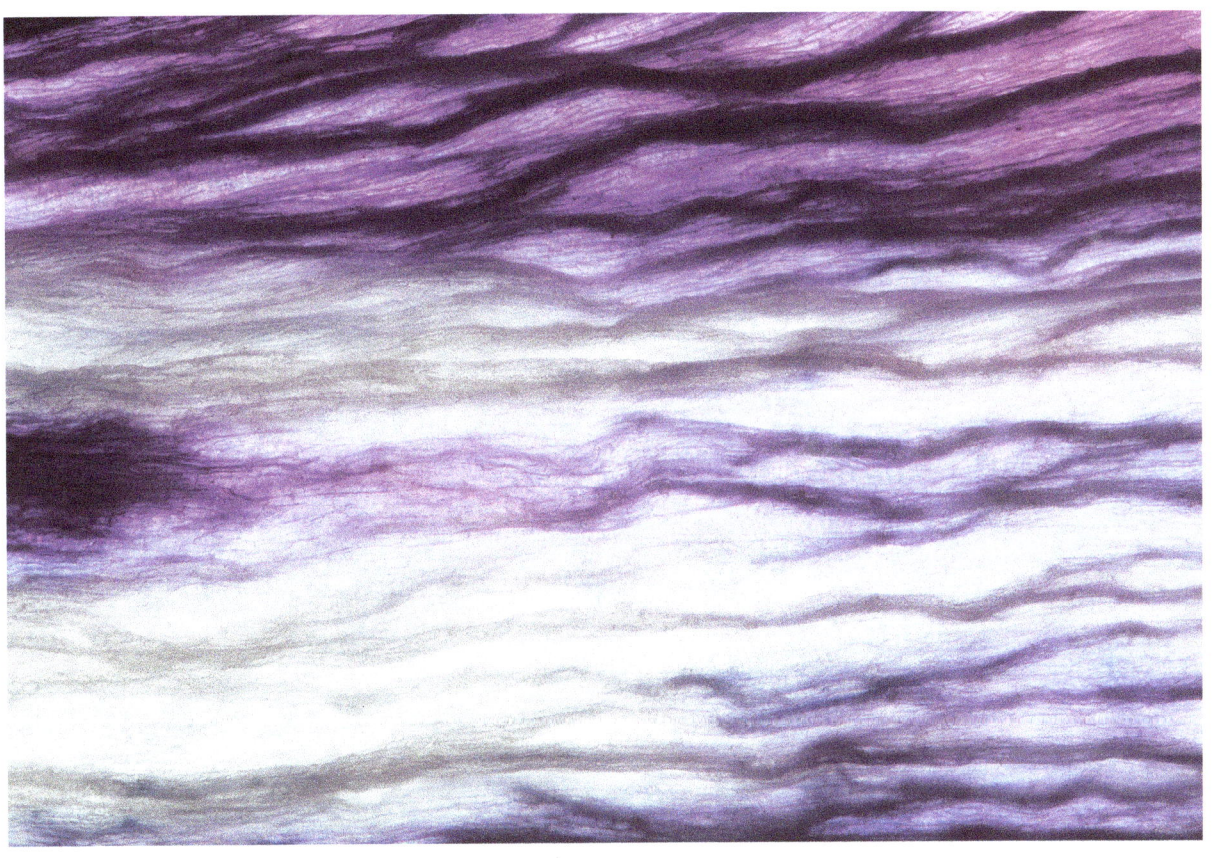

A photo of a tulip (*Tulipa*) petal taken with two lenses as one lens: MC Jupiter-37A 135mm f/3,5, which was made in 1985, and Zenitar-M 1,7/50, made in 1982. Magnification is 2.5x.

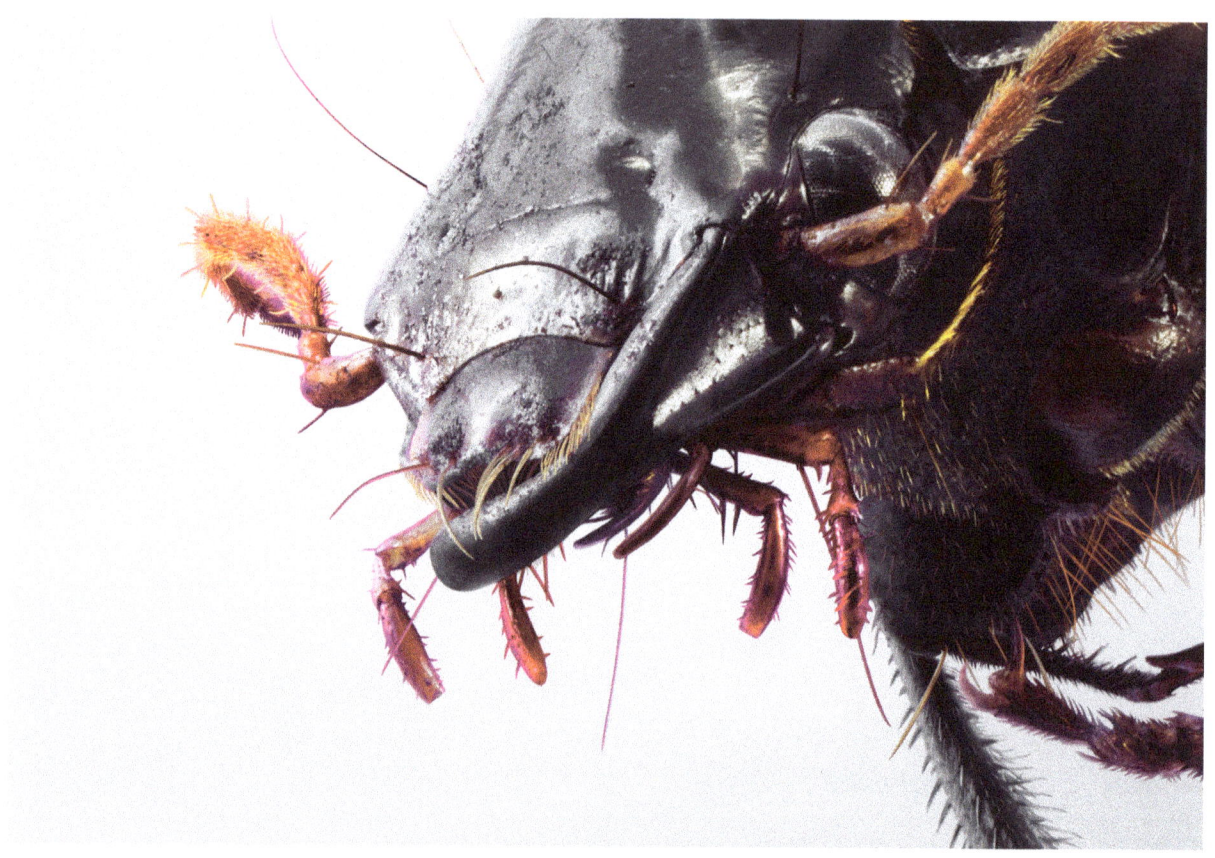

A microcosmic portrait of a com ground beetle (*Zabrus gibbus*) taken with two lenses as one lens: SMC Takumar 200mm f/4, which was made in 1982, and Zenitar-M 1,7/50, made in 1982. This image was stacked from many shots into one very sharp photo. Magnification is 4x.

Juxtaposition

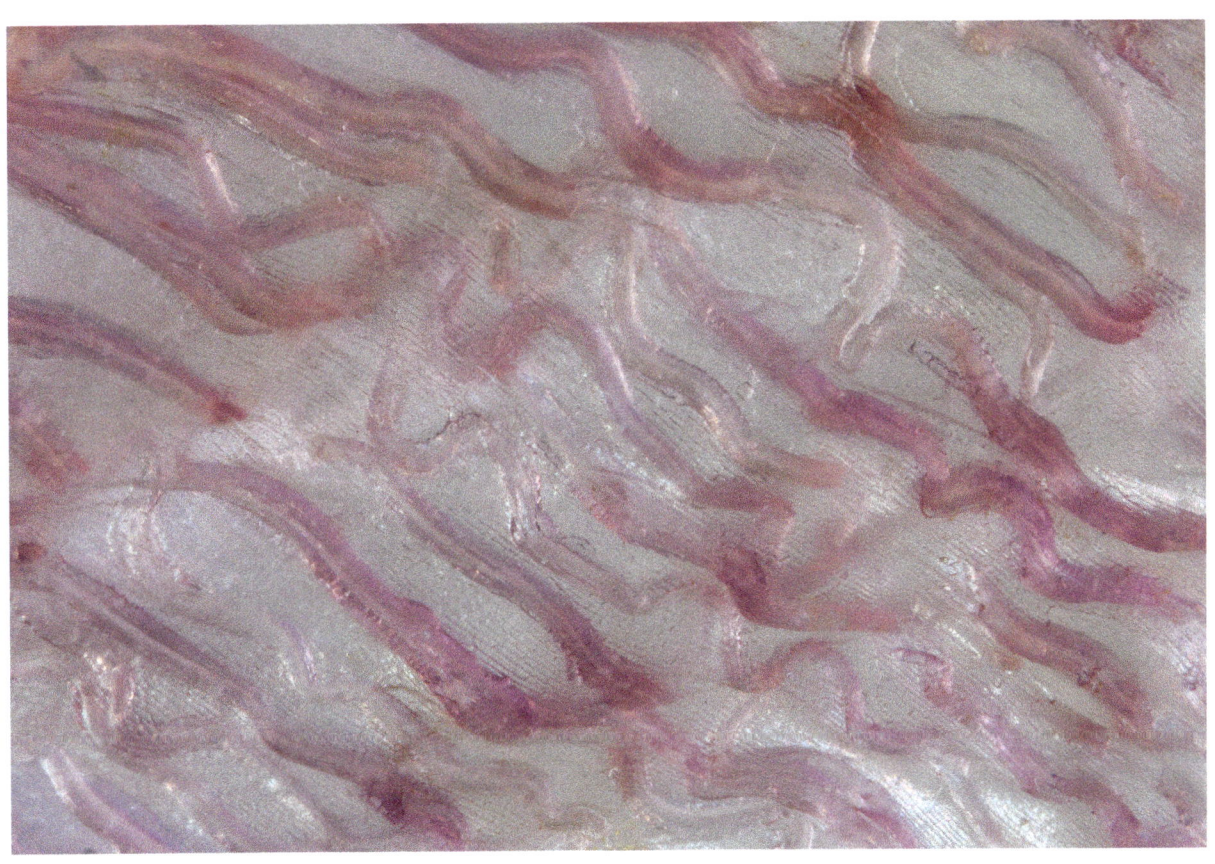

A photo of a tulip (*Tulipa*) petal taken with two lenses as one lens: MC Jupiter-37A 135mm f/3,5, which was made in 1985, and Zenitar-M 1,7/50, made in 1982. Magnification is 2.5x.

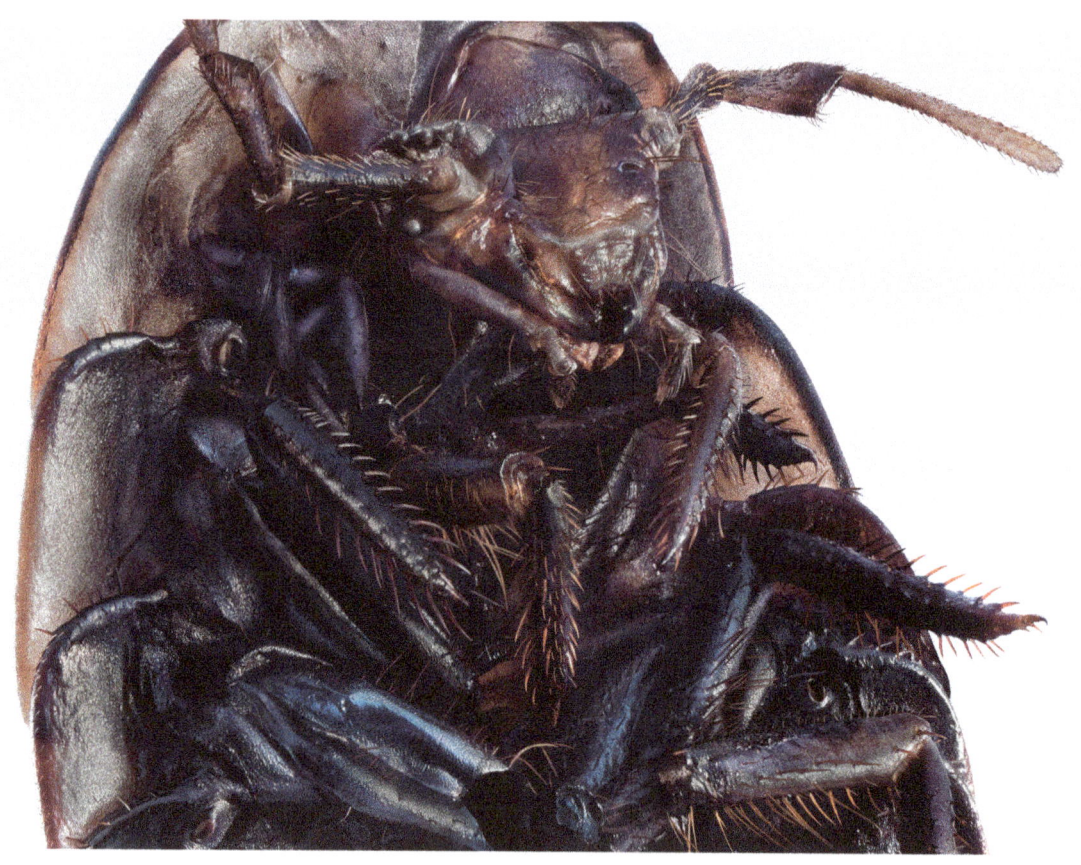

A microcosmic portrait of a woodlouse (*Oniscidea*) taken with two lenses as one lens: MC Jupiter-37A, which was made in 1985, and a retro lens Zenitar-M 1,7/50, made in 1982. This image was stacked from many shots into one very sharp photo. Magnification is 2.5x.

Juxtaposition

A photo of a tulip (*Tulipa*) petal taken with two lenses as one lens: MC Jupiter-37A 135mm f/3,5, which was made in 1985, and Zenitar-M 1,7/50, made in 1982. Magnification is 2.5x.

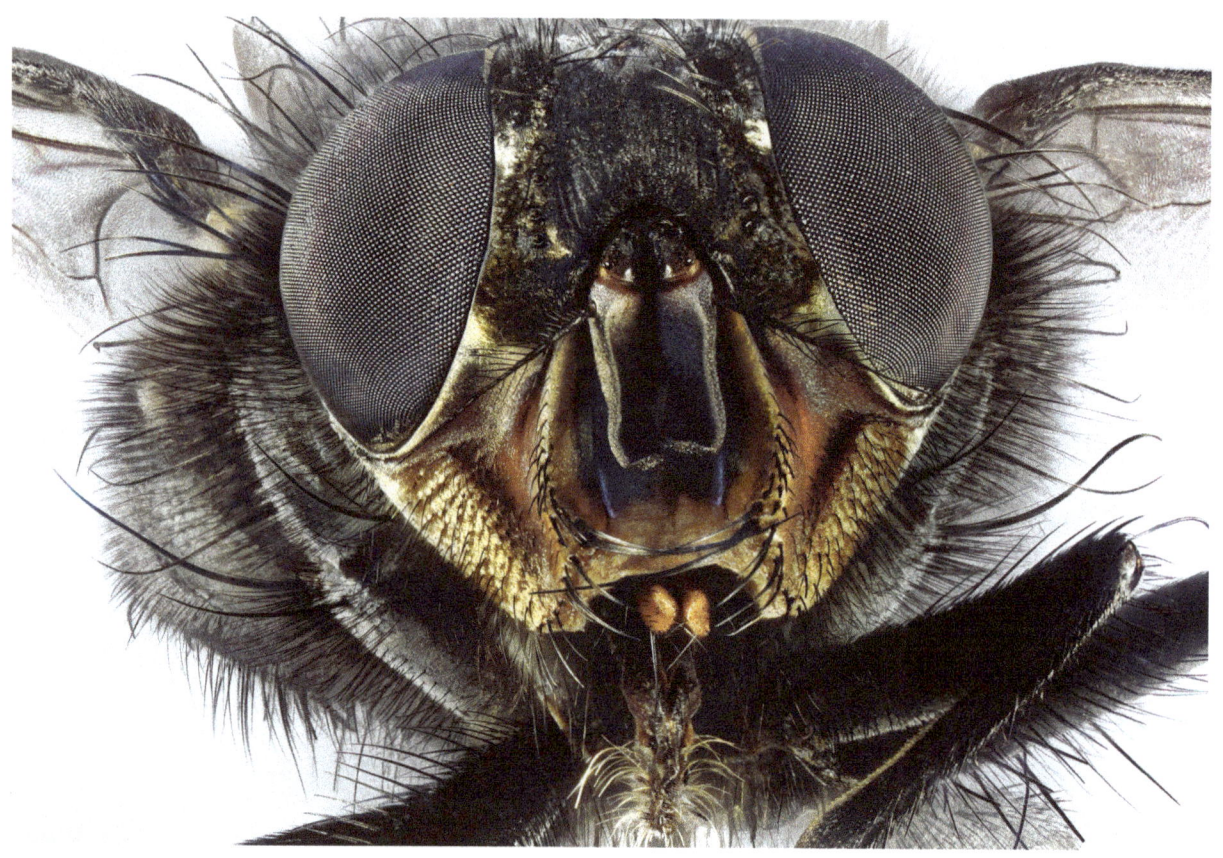

A microcosmic portrait of a housefly (*Musca domestica*) taken with two lenses as one lens: MC Jupiter-37A, which was made in 1985, and Zenitar-M 1,7/50, made in 1982. This image was stacked from many shots into one very sharp photo. Magnification is 2.5x.

Juxtaposition

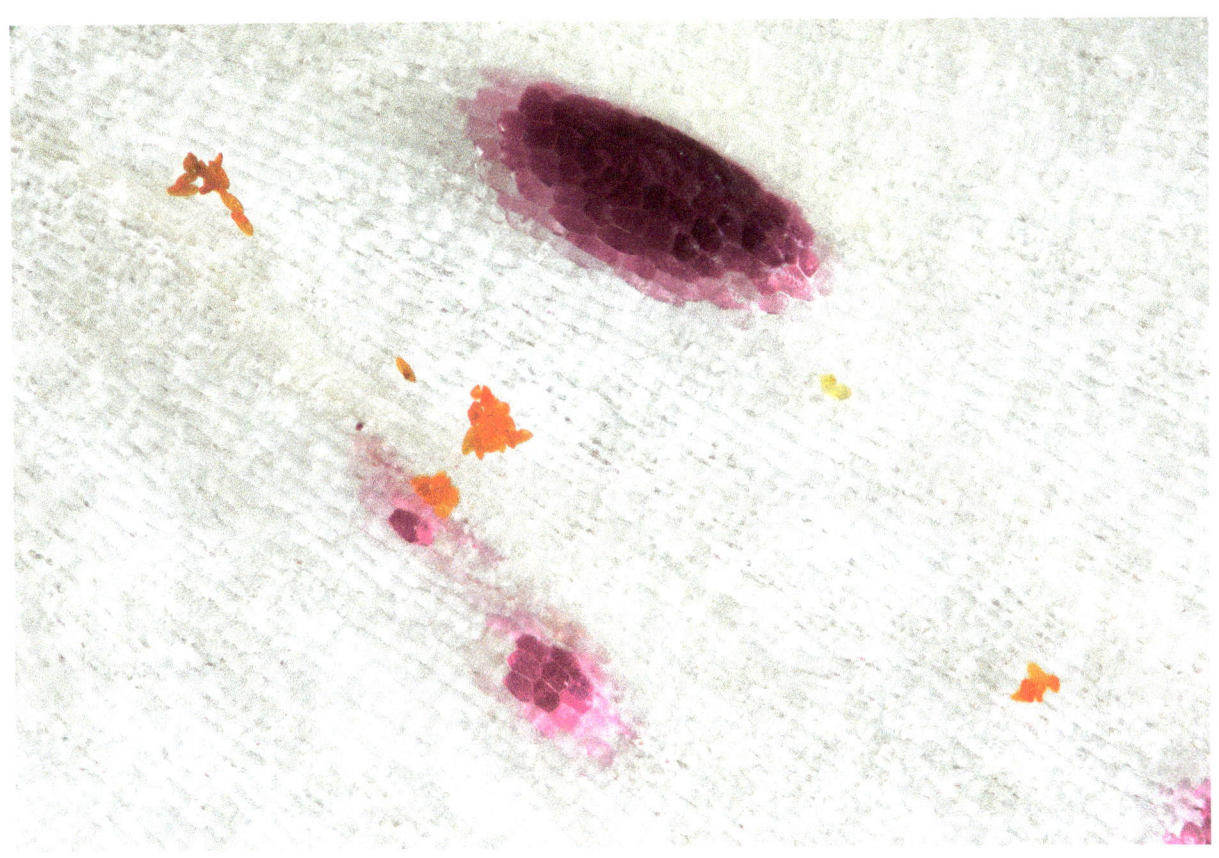

A micro photo of a lily (*Lilium*) petal taken with two lenses as one lens: SMC Takumar 200mm f/4, which was made in 1982, and Zenitar-M 1,7/50, made in 1982. This image was stacked from many shots into one very sharp photo. Magnification is 4x.

Irina Petrova Adamatzky

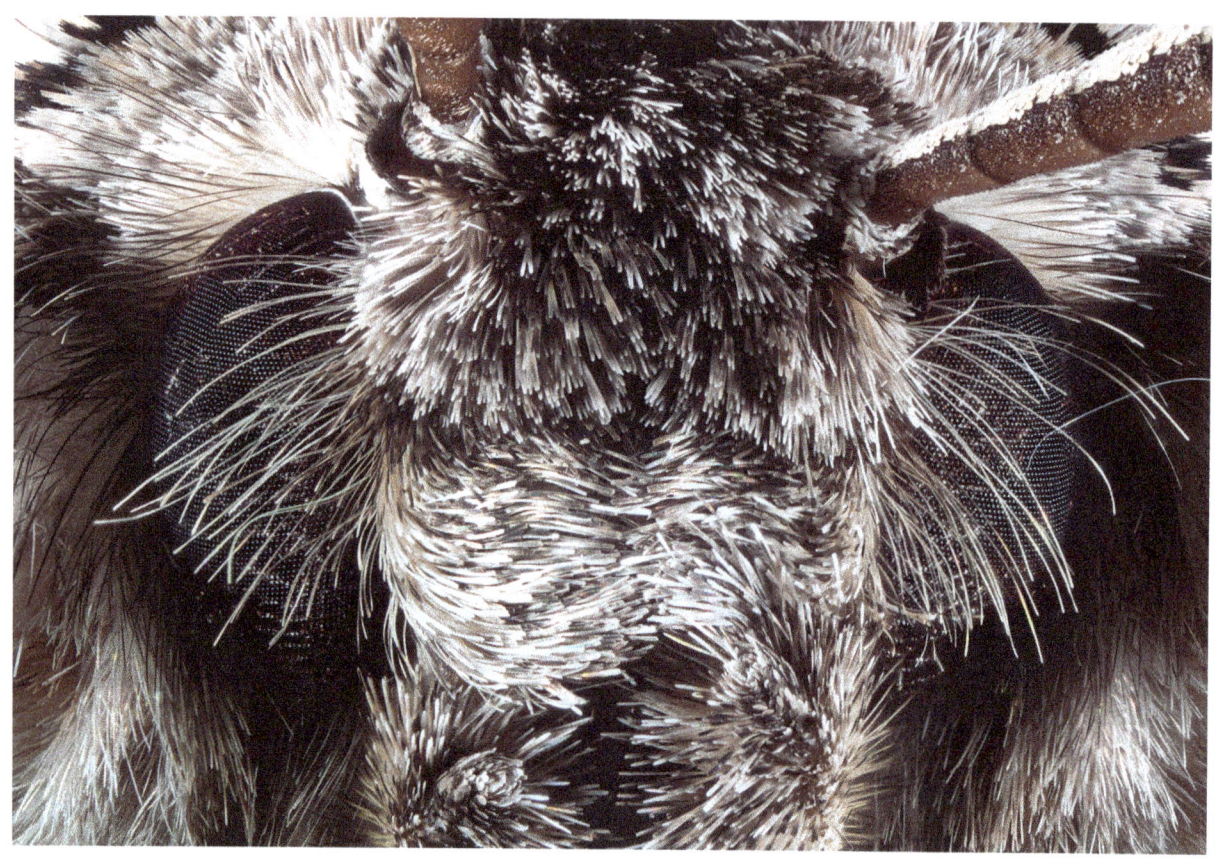

A microcosmic portrait of an owlet moth (*Noctuidae*) taken with two lenses as one lens: MC Jupiter-37A, which was made in 1985, and Zenitar-M 1,7/50, made in 1982. This image was stacked from 250 shots into one very sharp photo. Magnification is 2,5x.

Juxtaposition

Irina Petrova Adamatzky

Index

Arctia caja, 16,18,20,26,46
Aporia crataegi, 44,48
Betula, 21,23
Calliteara pudibunda, 14
Chironomidae, 54,56
Formicidae, 34,36
Gomphocerus sibiricus, 24
Iris, 45,49
Lilium, 15,25,29,31,35,37,51,57,65
Musca domestica, 30,64
Noctuidae, 66
Oniscidea, 62
Opiliones, 52
Pentatomidae, 38
Polistinae, 22,28,32,42
Rosa, 33
Solanum, 39,41
Stegobium paniceum, 58
Tabanidae, 50
Tipulidae, 40

Tulipa, 13,17,19,27,43,47,53,55,59,61,63
Zabrus gibbus, 60

Ant
    Formicidae, 34,36
Beetle
    Stegobium paniceum, 58
    Zabrus gibbus, 60
Birch
    Betula, 21,23
Bug
    Pentatomidae, 38
Butterfly
    Aporia crataegi, 44,48
Caterpillar
    Arctia caja, 20,26
    Calliteara pudibunda, 14
    Garden tiger moth, 20,26
    Pale tussock, 14
Chironomid

Chironomidae, 54,56
Com ground beetle
   Zabrus gibbus, 60
Crane fly
   Tipulidae, 41
Daddy longlegs
   Opiliones, 52
Drugstore beetle
   Stegobium paniceum, 58
Exoskeleton
   Pentatomidae, 38
Gadfly
   Tabanidae, 50
Garden tiger moth
   Arctia caja, 16,18,20,26,46
Ground beetle
   Zabrus gibbus, 60
Harvester
   Opiliones, 52
Harvestman
   Opiliones, 52
Horsefly
   Tabanidae, 50
Housefly
   Musca domestica, 30,64
Iris petal
   Iris, 45,49
Lake fly
   Chironomidae, 54,56
Larva
   Pentatomidae, 38
Leaf
   Birch, 21,23
   Betula, 21,23
   Potato, 39,41
   Solanum, 39,41
Lily petal
   Lilium, 15,25,29,31,35,37,51,57,65
Locust
   Gomphocerus sibiricus, 24

Midge
    Chironomidae, 54,56
Mosquito
    Chironomidae, 54,56
Moth
    Arctia caja, 16,18,20,26,46
    Calliteara pudibunda, 14
    Garden tiger, 16,18,20,26,46
    Noctuidae, 66
    Owlet, 66
    Pale tussock, 14
Nonbiting midge
    Chironomidae, 54,56
Owlet moth
    Noctuidae, 66
Pale tussock
    Calliteara pudibunda, 14
Paper wasp
    Polistinae, 22,28,33,42
Petal

Iris, 45,49
Iris, 45,49
Lily, 15,25,29,31,35,37,51,57,65
Lilium, 15,25,29,31,35,37,51,57,65
Tulip, 13,17,19,27,43,47,53,55,59,61,
    63
Tulipa, 13,17,19,27,43,47,53,55,59,61,
    63
Rosa, 33
Rose, 33
Rose petal
    Rosa, 33
Potato leaf
    Solanum, 39,41
Shield bug
    Pentatomidae, 38
Stink bug
    Pentatomidae, 38
Tulip petal
    Tulipa, 13,17,19,27,43,47,53,55,59,61,

63

Wasp
 Polistinae, 22,28,32,42
Woodlouse
 Oniscidea, 62

www.ingramcontent.com/pod-product-compliance
Lightning Source LLC
Chambersburg PA
CBHW040128230526
45473CB00032B/3046